D1824025

INSTITUTE OF GEOLOGICAL SCIENCES

Natural Environment Research Council

OVERSEAS MEMOIR 6

The geology of Guadalcanal, Solomon Islands

B. D. Hackman

London   Her Majesty's Stationery Office   1980

© *Crown copyright 1980*

Author

B. D. Hackman, BSc, PhD, ARCS
Ministry of Natural Resources, Honiara, Solomon Islands
and *Institute of Geological Sciences, London*

Bibliographical reference

HACKMAN, B. D. – 1980.   The geology of Guadalcanal,
Solomon Islands.   *Overseas Mem. Inst. Geol. Sci.*, No. 6.

ISBN 0 11 884080 0*

CONTENTS

Abstract v
One Introduction 1
Geographic setting 1
Geological setting 1
Previous research 3
Working methods 7
  Maps and aerial photographs 7
  Field methods 7
General observations on Guadalcanal 8
  Communications 8
  Geographical names 9
  Population 9
  Diseases 9
  Agriculture 9
  Vegetation 9
  Fauna 9
  Climate 10
Geomorphology 10
  Physiographic zones 10
  Coastal features 12
  Drainage 15
  Landslides and lakes 16
  Areas of calcareous rock 17

Two Stratigraphy 17
The pre-Miocene basement 17
  The Mbirao Group 17
  Mbirao Volcanics 18
  Tetekanji Limestones 19
  Mbirao Dolerites 20
  Guadalcanal Gabbro 21
  Mbirao Metabasics 22
  Petrochemistry of the Mbirao Group 23
The Guadalcanal Ultrabasics 25
  Marau Ultrabasics 26
  Suta Ultrabasics 27
  Ghausava-Itina Ultrabasics 27
  The possibility of other ultrabasic areas 28
  Petrochemistry of the ultrabasics 28
The Neogene–Pleistocene sequence 28
  Suta Volcanics 28
  Poha Diorite 30
  Kavo Greywacke Beds 32
  Mbetilonga Group 33
  Tangareso Beds 35
  Mbokokimbo Formation 36
  Toni Formation 38
  Mbalisuna Gabbro 41
  Vatumbulu Beds 41
  Pliocene sediments in western Guadalcanal 42
  Pliocene–Quaternary igneous activity 43
    Gallego Lavas 43
    Koloula Diorite 44
    Geothermal Springs 46
  Honiara Beds and associated sediments 46
  Raised Pleistocene terraces 47
  Recent sediments 48

Three Structural geology of Guadalcanal 48
Bedding planes in the Mbirao Volcanics 48
Metamorphic s-surfaces 52
  The sequence of s-surfaces 54
  Significance of s-surfaces within the Mbirao Group 56
Bedding planes in the Neogene sequence 56
  Deposition of sediments in western Guadalcanal 58
Folding 59
  Minor folds within the Mbirao Group 59
  Minor folds within the Neogene succession 60
  Regional folding 60
Faulting 60
  Fault lineaments 60
  Major faults 61
  Strike-slip faulting 61
  Evidence for transcurrent movement 63
  Interpretation of fault-patterns 64
  Faulting in the Mbirao Group and Suta Volcanics 64
  Faulting in the sediments 66
  Pattern of faulting in western Guadalcanal 66
Jointing 66
Discordant minor intrusives 67
Interim tectonic synthesis 69

Four Economic geology of Guadalcanal 70
Deposits associated with the Mbirao Group 71
  Copper in oceanic tholeiites of the Mbirao Volcanics 71
  Massive sulphide deposits in the Valasi Block 73
  Manganese 73
  Gold in the Suta area 73
  Nickeliferous laterites 73
Deposits associated with the post-Eocene felsic igneous association 74
  Koloula Diorite 74
  Poha Diorite 74
  Mbetilonga 75
Low-grade disseminated sulphides associated with the Gallego Volcanics 75
Gold deposits 76
  Gold Ridge 76
  Alluvial gold 76
  Possibilities of gold in other areas 77
Fossil fuels 77
  Petroleum 77
  Lignite 77
Residual deposits 77
  Bauxite 77
  Iron 78
  Beach-sand prospects 78
  Cementstone prospects 78

Five Geophysical investigations 79
Gravity 79
  History of research 79
  The regional picture 79
  Gravity pattern in Guadalcanal 80

Isostatic compensation 81
Gravity trends 81
Conclusions 82
Seismicity 83
Magnetic survey 83
Lines of geophysical discontinuity 86

Six Regional geology 87
Summary of geological history 87
The oceanic phase 87
The tectonic phase 87
The later volcanic or fractured island-arc phase 88
Renewal of the tectonic phase 89
The shaping of Guadalcanal 89
Comparative regional geology 90
Pre-Miocene basement 90
Suta Volcanics and Poha Diorite 91
Kavo Greywacke Beds 91
Miocene calcareous sediments 91
Plio-Pleistocene sedimentary sequences 92
Plio-Pleistocene igneous activity 92
Structural geology 92
The Solomons Fractured Arc 93
Theories on Melanesian geology in relation to global tectonics 93
The evolution of geanticlinal welts 93
Petrochemical evolution 94
The significance of regional stress changes 99
Speculations on the Solomons Fractured Arc 100
Plate tectonics 101

References 102

Appendix List of topographic names and their co-ordinates 108

PLATES

1 View looking north-east from the summit of Sambahatangi 8
2 View looking eastwards down the Sutakiki Valley 10
3 Basaltic pillow lavas with interstitial epidote, quartz and calcite, Mbirao Volcanics 18
4 Photomicrograph of porphyroid chert, Mbirao Volcanics 19
5 Dolerite dyke, with feldspar glomerocrysts showing flow-alignment, in Mbirao Group 20
6 Photomicrograph of leucogabbro, Guadalcanal Gabbro 21
7 Photomicrograph of graphite-schist, Mbirao Metabasics 22
8 Photomicrograph of harzburgite, Marau Ultrabasics 26
9 Serpentised harzburgite with dykes of anorthosite and enstatite-leucogabbro, Suta Ultrabasics 28
10 Algal balls in a matrix of volcanic rudite, Suta Volcanics 29
11 Biomicrite with lithic fragments, Valasi Limestone 34
12 Photomicrograph of silty biomicrite, Tangareso Beds 36

13 Jointing in the Koloula Diorite 45
14 Gravity folding in Recent laminated lacustrine silts 47

FIGURES

1 Map of the Melanesian archipelagos vi
2 The geological provinces of the Solomon Islands 2
3 Stratigraphic correlation table for Guadalcanal 6
4 Schematic stratigraphic section across northern Guadalcanal *in pocket*
5 Map of physiographic zones in Guadalcanal 13
6 Topographic cross-sections of Guadalcanal 12
7 Map showing the location of geothermal springs in Guadalcanal 46
8 Tectonic domains of Guadalcanal 49
9 Contoured equal-area stereographic projection of poles to bedding planes in the Mbirao Group 52
10 Contoured equal-area stereographic projections of poles to metamorphic s-surfaces in the pre-Miocene of Guadalcanal 50
11 Maps showing trends of metamorphic s-surfaces in south-east Guadalcanal 54
12 Changes in the direction of maximum stress as deduced from patterns of s-surfaces in Guadalcanal 55
13 Contoured equal-area stereographic projections of poles to bedding planes in the post-Oligocene sediments of central and eastern Guadalcanal 56
14 Stereograms showing the plunge of fold axes in Guadalcanal 59
15 Diagrams showing the trends of major faults in Guadalcanal 62
16 Map showing the influence of the Venru Moli Fault on the pattern of drainage in south-eastern Guadalcanal 63
17 Rose-diagram showing the trends and character of transcurrent faults in Guadalcanal 64
18 Rose-diagrams showing the strike of faults in eastern and central Guadalcanal 65
19 Rose-diagrams showing the strike of joints in eastern and central Guadalcanal 65
20 Rose-diagrams showing the strike of discordant minor intrusives in Guadalcanal 67
21 Rose-diagrams showing the strike of mineralised veins in eastern and central Guadalcanal 71
22 Map of prospects for economic minerals in Guadalcanal 72
23 Map showing the trends of gravity anomalies in the Solomon Islands 79
24 Map showing Bouguer anomalies, and geophysical units and discontinuities deduced from magnetic, electromagnetic and gamma-ray spectrometric data 81
25 Map of seismic activity in the Solomons–Santa Cruz region 83
26 Map of the structural elements of the Melanesian arcs 94
27 Cross-sections illustrating the evolution of the geanticlinal welts of Guadalcanal and Malaita 95

28 Silica: alkalis variation diagram for igneous rocks of Guadalcanal 96

29 'F'MA diagram for igneous rocks of Guadalcanal and Savo 96

30 Von Wolff diagram for igneous rocks of Guadalcanal and Savo 97

31 Von Wolff diagram for igneous rocks of the Solomon and Santa Cruz Islands 98

32 Murata's variation diagram for igneous rocks of Guadalcanal 98

33 Variation of azimuth of maximum regional horizontal stress in relation to time 99

TABLES

1 The lithostratigraphic units in Guadalcanal described by various authors from 1955 to 1969 4

2 Chemical analyses of basalts from the Mbirao Group 23

3 Chemical analyses of schist, dolerites, microgabbro and leucogabbro from the Mbirao Group 24

4 Chemical analyses of rocks from the Guadalcanal Ultrabasics and their associates 25

5 Chemical analyses of rocks from the Suta Volcanics 31

6 Chemical analyses of Tertiary plutonic rocks 32

7 Chemical analyses of rocks from the Gallego Lavas 32

8 List of localities where sequences of metamorphic s-surfaces can be observed 53

9 The sequence of tectonic events in Guadalcanal 68

MAP *(in pocket)*

Geological map of Guadalcanal, Solomon Islands

ABSTRACT

The stratigraphy and structural geology of the island of Guadalcanal, Solomon Islands, are described in a measure of detail appropriate to geological mapping on the scale of 1:50 000.

The pre-Miocene basement of Guadalcanal comprises the Mbirao Group and the Guadalcanal Ultrabasics. The Mbirao Group has been subdivided into five new units: essentially a complex of 'oceanic' basalts, subsidiary pelagic limestones and minor cherts, it has been intruded by cognate dolerite sills and dykes and, in the Upper Cretaceous, by discordant bodies of leucocratic uralite-gabbro. The Mbirao Group was warped in the early Tertiary to form a major geanticlinal welt: Alpine-type ultrabasics were emplaced into the axial zone of the welt, and at depth the basic igneous rocks were metamorphosed to greenschist and amphibolite facies. A sequence of four sets of metamorphic s-surfaces has been recognised in the metabasic rocks, ranging in age from Eocene to Pliocene. The changes in trend of these s-surfaces are explained in terms of a progressive clockwise rotation of the maximum horizontal stress.

A sequence of up to 6000 m of volcanic and sedimentary rocks overlies the pre-Miocene basement: The Oligocene–Miocene Suta Volcanics, a pile of basaltic andesites, provided the source material for a succession of Miocene greywackes; together with the Miocene limestones of the Mbetilonga Group these formations were gently folded and faulted in the late Miocene.

The Plio-Pleistocene sequence of basaltic and andesitic volcanics, volcaniclastic arenites and rudites with subsidiary coralgal reef, shows very rapid facies changes; in central and western Guadalcanal the calcalkaline trend is manifested by the effusion of hornblende-andesite in zones of extensive fracturing and the emplacement of a dioritic stock. There is evidence for vertical uplift of at least 800 m relative to sea level since the end of the Pliocene and of over 2500 m since the Miocene.

An attempt is made to integrate the roles of vertical movement and transcurrent faulting, also to explain observed fracture patterns in terms of the history of structural events as recorded by the s-surfaces.

The regional geological history is discussed in the light of geophysical information, and compared with that of neighbouring areas; the Solomons fractured arc is conceived as having evolved from a series of oceanic welts which started to shoal in the Lower Miocene. Fracturing under the influence of deep-seated transcurrent faulting may have led to the rotation of discrete blocks, e.g. the Mbirao Block, on which was periodically impressed the effects of a regional north-east–south-west stress system.

The significance of the regional geological history is discussed in relation to current theories of plate tectonics.

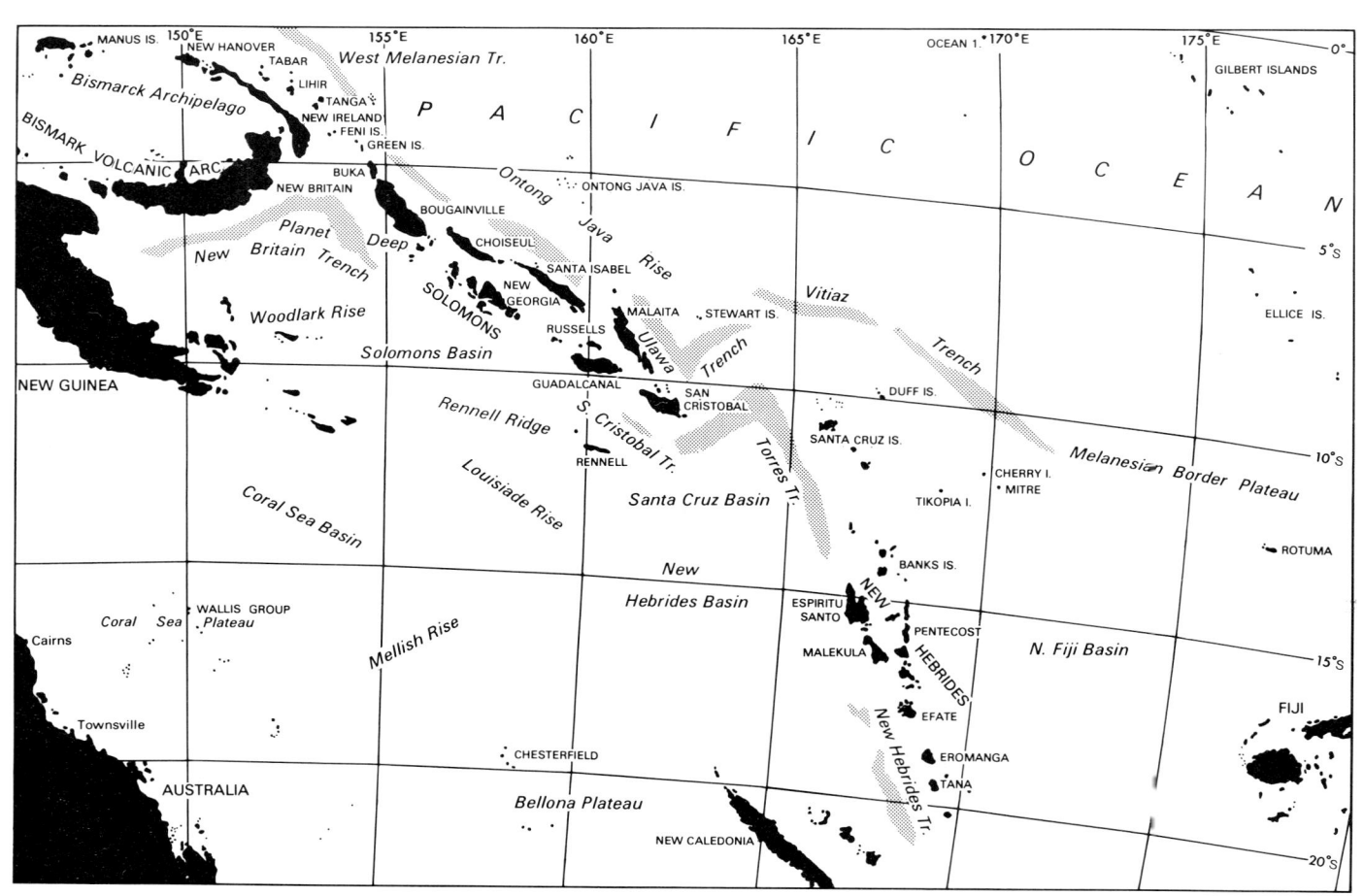

Figure 1 Map of the Melanesian archipelagos

ONE

Introduction

Geographic setting

Guadalcanal (about latitude 9°15′–10°S, longitude 159° 30′–160°E) is the largest of the six major islands of the Solomons group, approximately 6000 km² in area, being 150 km long from north-west to south-east and, at its broadest, 45 km wide. It is the centre island of the southern half of a double *en échelon* chain, that extends from Buka-Bougainville in the north-west, in the Australian territory of New Guinea, for 1200 km to beyond San Cristobal in the southeast. On geological grounds the Lihir and Tabar islands might be regarded as a north-westerly extension of the Solomons structure, continuing for a further 250 km along the north-east flank of New Ireland (Figure 1).

Guadalcanal lies south of the gap between Malaita and Santa Isabel in the northern chain, just as the New Georgia group is centred opposite the gap between Choiseul and Santa Isabel, enhancing the overall *en échelon* or rhombic symmetry of the double chain. Although politically the Solomon Islands do not include Buka and Bougainville, they do include the Santa Cruz group, which is bathymetrically a northern extension of the New Hebrides ridge.

The Solomons form an island chain whose axis trends at about 45° south of east; it might be considered as one segment of a complex of chains or island arcs and bathymetric highs extending from Fiji in the east through the New Hebrides to New Guinea and Indonesia in the west. The axis of the chain is very slightly convex towards the Coral Sea. The bathymetry of the south-western or Coral Sea side of the Solomons shows a linear trench which reaches a depth of over 6000 m, 80 km south of San Cristobal. The trench is shallower south of the New Georgia Group, but deepens again to 9140 m in the Planet Deep of the New Britain trench. This shallower portion of the trench occurs where a submarine ridge trending north-east, the Woodlark Rise, abuts onto the Solomons chain from the eastern end of the Papuan promontory. In a similar manner, a submarine ridge separating the Santa Cruz Basin from the New Hebrides Basin approaches the shallowest portion of the trench on the west side of the New Hebrides, between the New Hebrides and Torres trenches.

The limits of the Solomons segment are clearly defined by the sharp re-entrants in the trench system, to the north-west in the New Britain Trench, and to the south-east where the San Cristobal Trench meets the Torres Trench. These angles mark major changes in trend of the island chains—thus the Bismarck Volcanic arc is concave to the Coral Sea trending at about 60° east of north, with an overall east-west trend, while the axis of the Y-shaped New Hebridean chain bears 75° south of east, that is at an angle of about 30° to the axis of the Solomons.

Of the six major islands or island groups in the Solomons, namely Choiseul, Santa Isabel, Malaita, New Georgia, Guadalcanal and San Cristobal, the shape of Santa Isabel is the most obviously linear, whereas that of Guadalcanal is more stumpy with a sigmoid 'twist'. Between New Georgia and Santa Isabel, and between San Cristobal and Malaita, the sea depths exceed 2500 m, but elsewhere in between the two chains the depths are generally much less. In fact the entire area between Bougainville and San Cristobal forms a distinct rhomb-shaped bathymetric high with a slight sigmoidal irregularity, thus simulating the shape of Guadalcanal Island on a larger scale.

The Y-shaped configuration of the New Hebrides archipelago might be considered to be the southern half of a comparable, but more elongated rhomb-shaped bathymetric high which closes to the north in the Santa Cruz Group, at the eastern end of the Solomons.

The Russian bathymetric map of the Pacific Ocean (Tikhiy Okean, published in 1963) shows a less clearly defined oceanic trench to the north and east of the Solomons. The so-called Vitiaz Trench, named after a Soviet oceanographic vessel, extends from the northern side of the North Fiji Basin for 900 km in a north-westerly direction; locally this trench reaches a depth of over 7000 m. At the eastern end of the Solomons, this feature may be traced into a sharp re-entrant, the Ulawa Trench, which closely parallels the Torres–San Cristobal trench feature. To the north-west, flanking the islands from Malaita to Bougainville, this trench may be traced as a series of shallower basins between the Ontong Java Rise and the Solomons bathymetric high. The West Melanesian Trench, north of the Bismarck Archipelago, might be regarded as a more clearly defined north-westerly continuation of the same feature.

Geological setting

The best introduction to, and summary of, the geology of the Solomons is that of Coleman (1965a), who compiled a series of stratigraphical and structural notes to refer to the first Geological Map of the British Solomons published in 1962. Coleman (1970) describes the Solomons and New Hebrides within the regional framework of the south-west Pacific.

Figure 2 illustrates Coleman's subdivision of the islands into five geological provinces:
1 Central Province.
2 Pacific or Malaitan Province.
3 Volcanic or New Georgian Province.
4 Oceanic Volcanic Province.
5 Atoll Province.
The position of the boundaries of the Central Province has been amended somewhat from that of Coleman (1966) to accord with evidence recently available from Santa Isabel, in particular.

Most of Guadalcanal falls within the Central Province, although the north-western area represents the eastern end of the Volcanic Province. In addition, the Central Province embraces the whole of Choiseul, San Cristobal and Florida, together with the south-western side of Santa Isabel.

The islands of the Central Province are characterised by intensely faulted cores of pre-Miocene basic lavas, in part regionally metamorphosed to a low grade (greenschist or amphibolite facies). These are overlain by a sedimentary succession ranging in thickness from 5000 m in east-central Guadalcanal to perhaps less than 700 m on San Cristobal.

The sediments range in age from Lower Miocene through to Holocene, including organic limestones, calcarenite, arenaceous and volcaniclastic material. The sedimentary pile in general shows shallow dips, variations being due to block faulting or gentle 'drape' folding reflecting underlying basement structures.

The Pacific Province, including the whole of Malaita and the north-eastern flank of Santa Isabel, also has a core of basic lavas, although to date it has nowhere been found to be metamorphosed as in the Central Province. The overlying succession is of dominantly pelagic sediments; in contrast to the Central Province, it ranges in age from Cretaceous to Recent and is folded along north-west-trending axes.

Faulting also has occurred, but in north-central and southern Malaita basic lavas have been folded together with the overlying limestones; neither unconformity nor tectonic discordance has been recorded between the two divisions (according to Maranzana, 1968; also from the writer's own observations).

The Volcanic Province extends into north-west Guadalcanal, comprising cones and lava piles of basaltic and andesitic composition, surrounded by derived clastic sediments and fringing reef material.

The Oceanic Volcanic Province comprises the remoter Santa Cruz Group in the Eastern Solomons. Earlier work suggested that they were largely volcanic islands of the oceanic alkali-basaltic association, although recent investigations show that Santa Cruz also has a sedimentary sequence bearing some resemblances to that of the Central Province (Craig, in press), as well as to the sequences on parts of Santo and Malekula, New Hebrides (Coleman, 1970).

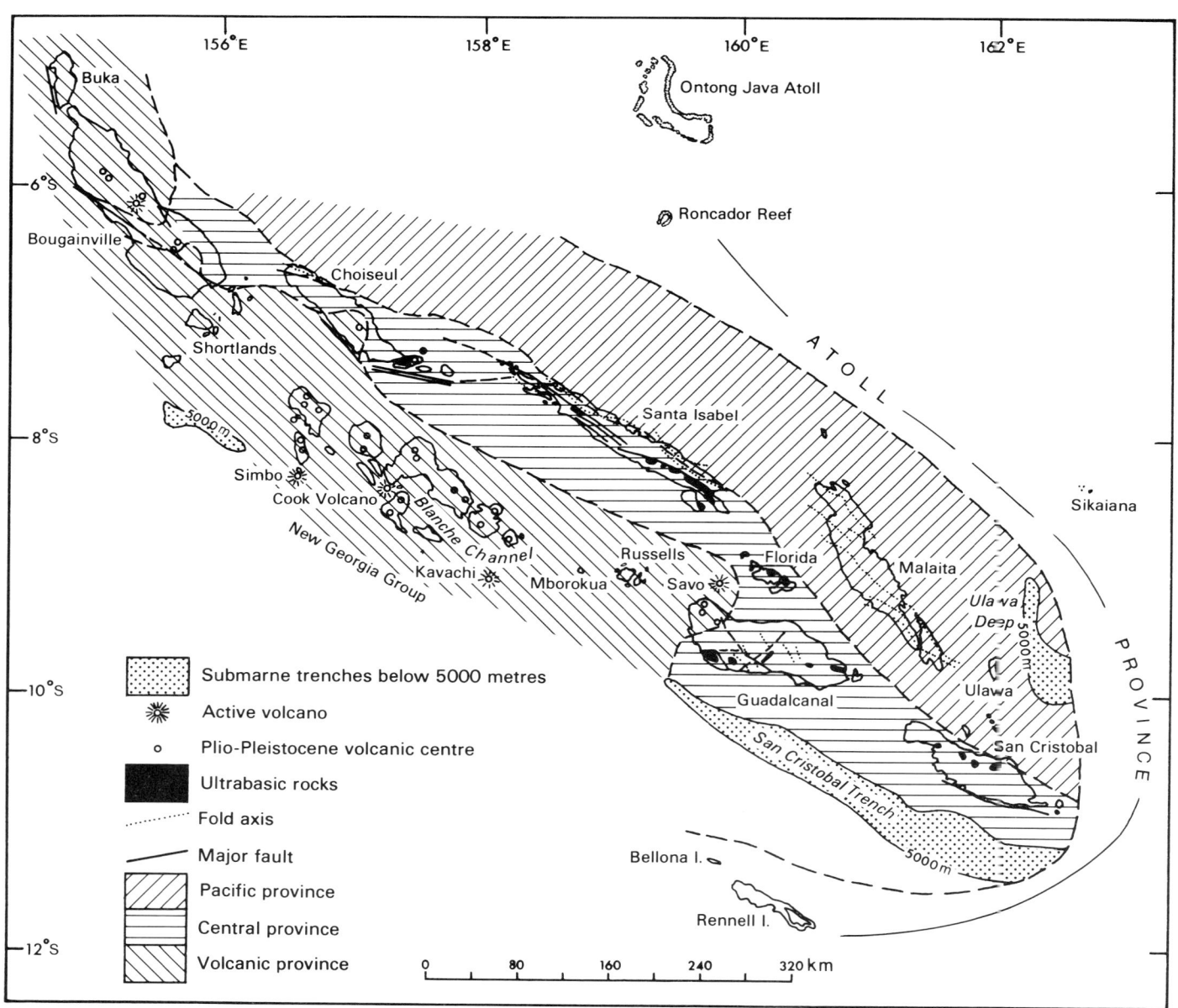

Figure 2 The geological provinces of the Solomon Islands

The Atoll Province includes a large number of outlying reefs and sand cays, in part uplifted. Nothing is known of their pre-Quaternary history.

Previous research

Guadalcanal was first visited by the Spaniard Alvaro de Mendaña in 1568 during his voyage of discovery to the Solomon Islands (Amherst and Thomson, 1901). The party landed near the mouth of the Matepono River; they were convinced that gold was present in the alluvial sand of that river, but were unable to explore extensively on account of the hostility of the natives.

Traders began regular visits to Guadalcanal during the late 19th century, but even in 1881 Guppy was unable to land on the island; he did however make some interesting observations from the boat (Guppy, 1887), remarking particularly on the contrast in surface configuration between the eastern and western parts of the island. He discovered boulders of 'schistose diabases, diorites and other dense basic rocks' which had been transported by floating trees to the small island of Rua Sura, off the north coast of Guadalcanal. He also observed that 'the mountainous ridges of the elevated eastern region so closely resemble the long level ridges of the interior of San Christoval, that I cannot doubt but that they are similarly formed of very ancient and often highly crystalline volcanic rocks'.

The first scientific expedition to penetrate into the interior of Guadalcanal was the ill-fated Austrian 'Albatros' expedition of 1896, led by von Norbeeck. Grover (1955c) gave a graphic description of their abortive attempt to climb Mt Tatuve.

In the same volume Grover detailed the history of the pre-war prospecting phase initiated when gold was discovered in the Sutakama and Matepono headwaters by the Australian botanist Kajewski.

In April, 1950, the British Solomon Islands Geological Survey was inaugurated with the arrival of Grover, and in December, 1950, Coleman, as a member of a University of Sydney expedition led by Professor Marshall, commenced investigations in north-western Guadalcanal. Partly because of proximity to Honiara, the capital, this part of the island and later north-central Guadalcanal, received more attention than the less accessible areas of the island.

The main stratigraphic units, based on reconnaissance work between 1950 and 1965, are summarised in Table 1. Priorities in nomenclature can be derived at once from this table; a multiplicity of local informal names has arisen, not all of which have been adopted by currently working geologists.

Pudsey-Dawson and Thompson (1958) followed up Coleman's early work in western Guadalcanal, somewhat modifying his 1957 section on the basis of more detailed surveying and petrographic work (Column 3). About the same time Grover and Pudsey-Dawson (1958) made the first traverses across eastern Guadalcanal (Column 4). Grover (1958a and b) pursued more detailed investigations in areas of possible economic interest, particularly Gold Ridge.

Coleman extended his investigations eastwards into north-central Guadalcanal; he produced a geological sketch map of the area between the Mataniko and Aola rivers, and detailed the Neogene succession (Coleman, 1960b; Table 1, Column 5), which differed in many respects from that established for western Guadalcanal.

Hill (1960) re-explored the Mbetilonga area, mapping in some detail up to the northern slopes of the Kavo Ranges. He accepted Coleman's 1960 succession for the west-central area, but introduced a new term, the Kavo Shales (and Siltstones), for a greywacke-type formation overlying the basement in certain of the head tributaries of the Lungga River, and believed to be in part Lower Miocene.

Connolly (1960) published the first observations on the Cape Henslow–Oa area in south-east Guadalcanal; he rejected the possibility of the existence of an economic iron-ore formation which had been reported by a wartime American Army reconnaissance party. Notes on Connolly's assessment of Guadalcanal beach sands are contained in Grover (1960).

Thompson (1960) investigated the ultrabasic rocks of two areas in Guadalcanal, Marau in the east and Ghausava in the west; his local successions for the two areas are given in Columns 7 and 8 of Table 1.

Coleman (1965a) presented his first synthesis of the geology of the Solomons, to accompany the First Geological Map, 1962; he introduced some minor amendments to the successions established for north-central Guadalcanal, adopting the name Kavo Greywacke Beds for the Kavo Shales of Hill (1960).

Wells (1965) and Stanton (1965) discussed the Recent gravels of the Kavahambe (Kovagombi) area of the lower Chovohio (Sorvohio) River with reference to the gold potential. Grover (1965) described several mineral occurrences of possible economic interest, mostly from western and central Guadalcanal, in which first mention was made of the occurrence of granodiorite in the Koloula River in south-central Guadalcanal; this report was based on unpublished field notes made by Pudsey-Dawson on a brief visit to Koloula in 1958.

Increased activity since 1962 Systematic geological mapping of Guadalcanal on the scale of 1:50 000 has been in progress since 1962.

Wright (1968a and b) was responsible for mapping the basins of the Lungga, Tenaru and Ngalimbiu river systems in west-central Guadalcanal. He endeavoured as far as was practicable to fit the mappable formations into the framework of Coleman's successions (1957 and 1960), but encountered considerable difficulties due to the rapid horizontal facies changes.

The geology of the Itina River Basin, to the south, was described by Thompson (1968); the sedimentary succession there appeared to have much in common with the Lungga Beds of Wright (1968b).

Research on foraminifera continued on samples from north-central Guadalcanal and elsewhere in the Solomons. Coleman and McTavish (1964) drew attention to the association of planktonic and larger benthonic foraminifera in the Middle Miocene sediments (Charikangge Beds) of the Tangareso (Tangaraisu) River, which has become a classic section in the history of Guadalcanal stratigraphy.

The period since 1964 was noteworthy for a tremendous increase in the amount of geophysical data obtained throughout the Solomon Islands. In 1963–1964 a land gravity survey of the major islands was conducted by a team from the University of Wisconsin, with the assistance of the US Army Map Service. Laudon (1968) gives a description of the project; some of the geological implications are discussed by Grover (1968b).

In the ensuing years (1966–1968) marine gravity and magnetic studies of the surrounding seas were made by a team from the University of Hawaii, under the direction of Woollard and Rose (Rose and others, 1968); also the sparker profiles from seismic reflection investigations have been made available in Woollard and others (1967).

The United Nations-sponsored Airborne Geophysical Survey was inaugurated in 1965; all the major islands were flown for magnetometric, electromagnetic and radiometric surveys (ABEM, 1967), although the mountainous terrain of south-central and south-eastern Guadalcanal had to be excluded from the survey.

The United Nations project made the following contributions to the geology of Guadalcanal:
1   The first map and description of the Koloula Granodiorite Complex, with reference to copper prospects (Winkler, 1968a).
2   The description of four exploration prospects of minor interest, at Gold Ridge, Mbetilonga, the Poha River and in the Talise area (Maranzana, 1968).
3   A photogeological interpretation on the scale of 1:50 000 by Allum (1967).

Since 1970 the geological mapping of western Guadalcanal has been completed on the scale of 1:50 000. The writer has been responsible for the Honiara and Tiaro Bay areas; the Cape Esperance–Ndoma area has been mapped by R. A. Dennis, the Lungga and Itina valleys by G. W. Hughes, and the Beaufort Bay area by C. C. Turner. The results of this work are unpublished to date.

Table 1   The lithostratigraphic units in Guadalcanal described by various authors from 1955 to 1969

Grover (1955b)	Coleman (1957)	Pudsey-Dawson and Thompson (1958)	Grover and Pudsey-Dawson (1958)	Coleman (1960b)	
1 Cent. Guadalcanal	2 W Guadalcanal	3 W Guadalcanal	4 E Guadalcanal	5 W Cent. Guadalcanal	6 E Cent. Guadalcanal
	Honiara Beds (Pleistocene–Recent)	*Honiara Beds* (Quaternary)	Recent alluvium and coastal limestone	*Ngalimbiu Alluvials* Honiara Beds	*Ngalimbiu Alluvials*
Kainozoic *Mudstone Formation* and intermediate eruptives	?U. Pliocene *Umasani Tuff*	?Pliocene *Umasani Tuffs* a few metres thick	?Pliocene mudstones and conglomerates	*Mt Austen Beds* (700 m)	Pliocene *Mt Austen Beds* (1000 m)
Basic eruptives	Pliocene *Lunga Volcanics* *Metanakau Tuffs* (Pliocene)	*Gallego Lavas* *Lunga Volcanics* including *Mataniko Tuffs*		*Tenaru Conglomerates* *Metapona Beds* (L.–U. Miocene)	
	Bonegi Limestone (L.–M. Miocene)	*Bonegi Limestone* (Miocene)	Limestone (?U. Miocene)	*Toni Tuffs* (650 m +) *Tangaraisu Marl* *Tina Calcarenite* (100 m) *Betilonga Limestone* (200 m)	*Berande Beds* (at least 1000 m) *Lake Lee Calcarenite*
Sutakiki Series metamorphics Granodiorite, quartz-diorite	*Guadalcanal Igneous Complex*	*Ultrabasic Intrusions* *Guadalcanal Igneous Complex* (Younger andesites and diorites; older andesites 450 m +)	Intrusive basic and ultrabasic rocks; ancient sediments and igneous rocks, folded and metamorphosed	*Basement Complex*	*Basement Complex*

R. D. Walshaw (1973) was responsible for a reassessment of the gold prospect in north-central Guadalcanal.

Structural geology Grover (1958a) attempted a preliminary regional synthesis of the structure of the central Solomons, remarking in particular on the occurrence of block faulting on Guadalcanal; he was impressed by the rhombic 'fish-net' pattern displayed by the major faults.

Coleman (1960b) stressed the importance of faulting as the dominant structural expression, there being no folding apart from 'minor undulations'. To quote: 'the structural picture, in summary, is one which shows a thick wedge-like cover of sediments resting unconformably on heavily faulted basement igneous rocks, the faults for the most part vertical or high-angle: the overlying sediments reflect movement in the basement blocks'.

With the exception of Thompson (1960) who recorded the local orientation of schistosities in the Marau and Ghausava ultrabasic areas, no attempt has been made hitherto to analyse mesoscopic structural data on a systematic basis.

In spite of considerable efforts by Coleman (1965a) to synthesise the geology of Guadalcanal, the impression gained from the previous literature is that stratigraphic information derived from separated areas by different authors has led to a multiplicity of informal local stratigraphic names (Table 1).

This was probably inevitable, considering the stage of reconnaissance mapping reached by 1962.

Much of the confusion must also be attributed to rapid horizontal facies changes; in fact nothing short of a painstaking survey of all the river sections is requisite, if the complexities are to be unravelled.

Thompson (1960)		Coleman (1965a)		De Golyer and Macnaughton (1965)	Hackman (1968c)
7 Ghausava area	8 Marau area	9 W Cent. Guadalcanal	10 E Cent. Guadalcanal	11 Cent. Guadalcanal	12 E Guadalcanal
Alluvium	Coral, alluvium and Recent beach deposits	*Ngalimbiu Alluvials* Honiara Beds	*Ngalimbiu Alluvials*	Sub-Recent deposits (200 m +)	Coral reef
	Pliocene *Berande Beds* (2000 m +)	*Mt Austen Beds*	*Berande Beds*	Upper sandy and conglomerate section (1500 m)	*Longgu Beds* (60 m) *Mbota Moli Beds* (120 m) *Vatumbulu Beds* (600 m)
Gausava Ultrabasics and Gabbro	*Marau Ultrabasics* (M. or U. Miocene)	[*Charikange Beds*] (Coleman and McTavish, 1964)	*Bokokimbo Beds* (M. and U. Miocene)	Middle marly/ silty section (?1000 m)	*Kolohaisava Mudstones* (230 m)
Marasa Tuffs	*Lake Lee Calcarenite* (?M. Miocene)	*Tangaraisu Shale Tina Calcarenite Betilonga Limestone* and *Kavo Greywacke Beds* (L. Miocene— Aquitanian)	*Lake Lee Calcarenite* (L. Miocene— Burdigalian)	Lower limy section (300 m)	*Isipeni/Valasi Limestones*
Guadalcanal Basement	*Kaukau Lavas* Intrusive gabbros *Guadalcanal Schists*	*Basement Complex*	*Basement Complex*	'Basement' diorites, andesites and altered basalts	*Mbirao Lavas* (1650 m +) with intrusive gabbros, metamorphosed in part

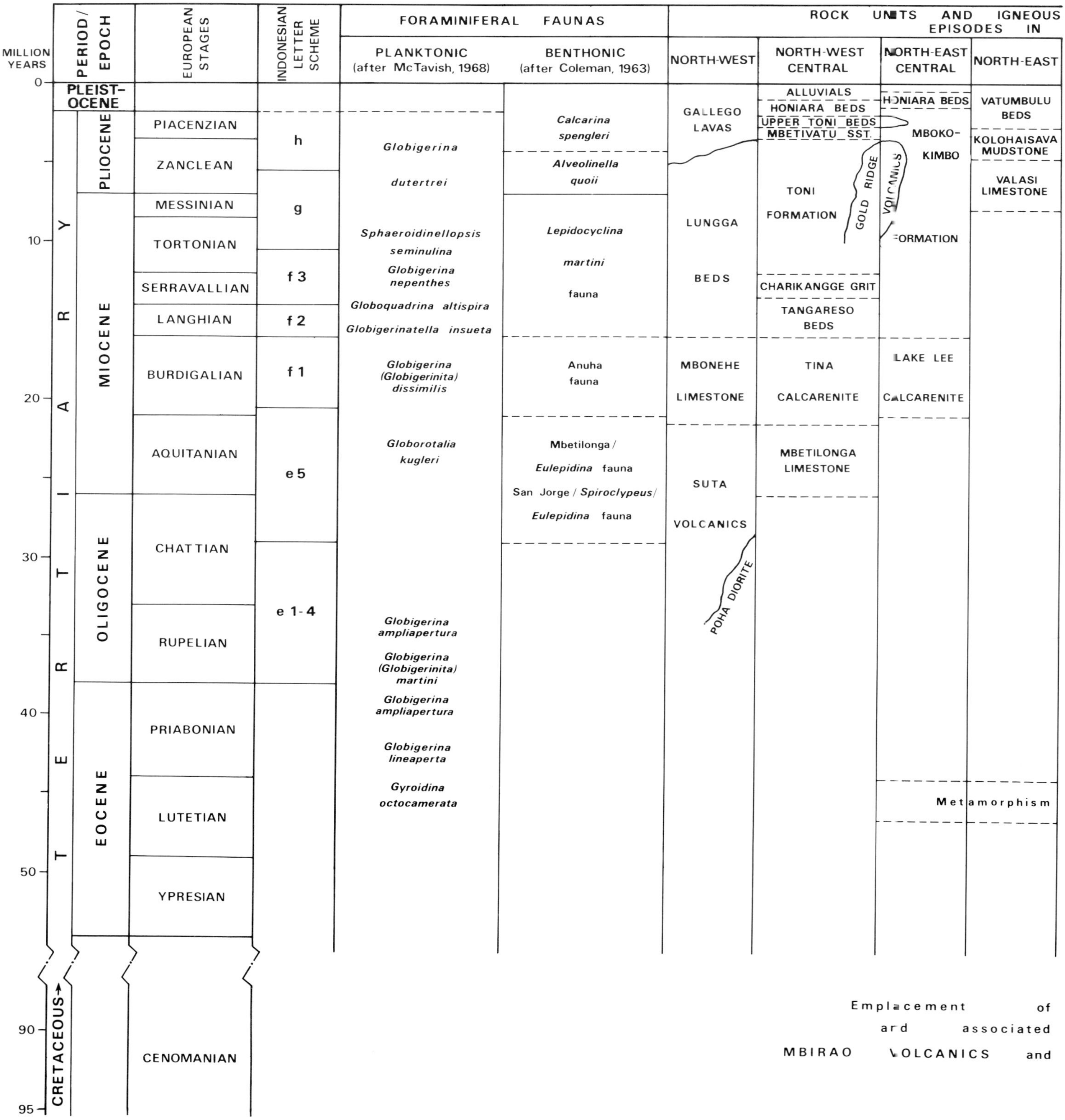
MILLION YEARS
PERIOD/ EPOCH
EUROPEAN STAGES
INDONESIAN LETTER SCHEME
FORAMINIFERAL FAUNAS
PLANKTONIC (after McTavish, 1968)
BENTHONIC (after Coleman, 1963)
ROCK UNITS AND IGNEOUS EPISODES IN
NORTH-WEST
NORTH-WEST CENTRAL
NORTH-EAST CENTRAL
NORTH-EAST
0
10
20
30
40
50
90
95
PLEIST-OCENE
TERTIARY
PLIOCENE
MIOCENE
OLIGOCENE
EOCENE
CRETACEOUS
PIACENZIAN
ZANCLEAN
MESSINIAN
TORTONIAN
SERRAVALLIAN
LANGHIAN
BURDIGALIAN
AQUITANIAN
CHATTIAN
RUPELIAN
PRIABONIAN
LUTETIAN
YPRESIAN
CENOMANIAN
h
g
f 3
f 2
f 1
e 5
e 1-4
Globigerina dutertrei
Sphaeroidinellopsis seminulina
Globigerina nepenthes
Globoquadrina altispira
Globigerinatella insueta
Globigerina (Globigerinita) dissimilis
Globorotalia kugleri
Globigerina ampliapertura
Globigerina (Globigerinita) martini
Globigerina ampliapertura
Globigerina lineaperta
Gyroidina octocamerata
Calcarina spengleri
Alveolinella quoii
Lepidocyclina martini fauna
Anuha fauna
Mbetilonga / Eulepidina fauna
San Jorge / Spiroclypeus / Eulepidina fauna
GALLEGO LAVAS
LUNGGA BEDS
MBONEHE LIMESTONE
SUTA VOLCANICS
POHA DIORITE
ALLUVIALS
HONIARA BEDS
UPPER TONI BEDS
MBETIVATU SST.
TONI FORMATION
GOLD RIDGE VOLCANICS
CHARIKANGGE GRIT
TANGARESO BEDS
TINA CALCARENITE
MBETILONGA LIMESTONE
HONIARA BEDS
MBOKO-KIMBO FORMATION
LAKE LEE CALCARENITE
VATUMBULU BEDS
KOLOHAISAVA MUDSTONE
VALASI LIMESTONE
Metamorphism
Emplacement of MBIRAO VOLCANICS and associated

Figure 3 Stratigraphic correlation table for Guadalcanal

AND METAMORPHIC GUADALCANAL			PHASES OF TECTONISM
SOUTH-WEST	SOUTH-CENTRAL	SOUTH-EAST	
		MBOTA MOLI BEDS	
	Emplacement of KOLOULA	GALLEGO LAVAS	
			PHASE 4
LUNGGA	DIORITE		
BEDS			PHASE 3
	KAVO GREYWACKE BEDS		
KAVO GREYWACKE BEDS MBETILONGA LIMESTONE MARASA VOLCANICS	SUTA VOLCANICS		
Some remobilisation of ultrabasics			PHASE 2
of Mbirao Metabasics			
GHAUSAVA ULTRABASICS	SUTA ULTRABASICS	MARAU ULTRABASICS	PHASE 1

Guadalcanal Gabbro

hypabyssal rocks

TETEKANJI LIMESTONES

General comments Only 27 chemical analyses are presented for rocks from Guadalcanal; hence it follows that conclusions drawn concerning the petrogenesis of the igneous rock types must be regarded as preliminary and tentative.

Coleman (1963, 1965b) and Coleman and McTavish (1964, 1967) have constructed a stratigraphic framework for the subdivision of the Neogene sediments within the Solomon Islands and New Hebrides, based on assemblages of larger benthonic and planktonic foraminifera. The writer has not attempted to enlarge that framework; rather, with the assistance of Dr Coleman and Dr G. W. Hughes, attention was given to foraminiferal samples and thin sections from crucial areas in order to extend the stratigraphic control over a wider area (Figure 3).

Sufficient of Guadalcanal has been mapped to enable reasonable generalisations to be made about the whole island; however, the writer has inevitably been biased in favour of eastern and central Guadalcanal. Parts of western Guadalcanal have been mapped in comparable detail by Thompson and Wright, but knowledge of certain areas is still sketchy.

The writer has endeavoured to scale the amount of descriptive detail to a standard consonant with that appropriate to the scale of mapping, that is to maintain an approach intermediate between a reconnaissance and a detailed survey.

Working methods

Maps and aerial photographs Tracings from the American Military Service (AMS) sheets, series X713, were used as base maps for plotting data in the course of field work. These maps are contoured at intervals of 20 m. Two sets of aerial photographs are available for most of Guadalcanal: the earlier USAF series (1942) and the RAF coverage of 1963.

Prior to the investigation of a selected area in the field, a photogeological interpretation was made to outline the major discernible contrasts, sketch in the lineaments, obtain supplementary topographical information and to assist in formulating a plan of action. This interpretation was revised in the light of what was discovered on the ground; lithological boundaries were completed on the map only as part of a final photogeological interpretation: in view of the densely forested nature of the country and the relative paucity of exposures available on the ridges and interfluves, this proved to be the most practicable method of completing the geological map. In very few places are climbs to the mountain summits rewarded with extensive bird's-eye views. An exception is the moss-forest plateau of the Kavo–Haiacha Ranges (Plate 1).

Field methods In the course of 56 weeks in the field, all the main rivers and major tributaries of central and eastern Guadalcanal were surveyed in detail, also a large proportion of the minor watercourses, to cover an area of approximately 3000 km². With the exception of the islands of the Marau archipelago at the eastern end of Guadalcanal, the writer mapped all that half of the island lying east of a line drawn from Mberande on the north coast to Nduindui on the south. In addition, check traverses were made into the headwaters of the Ngalimbiu River (previously mapped by Wright, 1968a).

The Honiara and Tiaro Bay areas were mapped in 1973 and 1974.

The river traverses were surveyed by a rapid pace-and-compass method, using two Solomon Islander geological assistants and a team of 12 to 16 native labourers. Except along the middle courses of major rivers, or in the proximity of easily identifiable geographic features, it was not found possible to locate oneself in the field on either maps or aerial photographs with a consistently high degree of accuracy. For this reason geological observations were plotted onto the field traverses on a scale of 1 cm to 100 paces. This information was then transferred selectively onto 1:50 000 compilation sheets which had been traced from the AMS map series.

Most of the rivers provide exposures along the greater part of their upper and middle courses, and a very close network of traverses can be obtained, although progress up the boulder-strewn creeks can be very slow and tedious. However, in some of the uppermost sections of the mountain courses, even the gorge outcrops can be obscured by thick growths of moss and lichen. In the lower middle courses of the larger rivers, exposures are usually limited to the laterally eroding bluffs on the convex sides of meanders, and are frequently deeply weathered; better sections can usually be found in the tributaries, particularly in the northern foothills zone.

Traverses on ridges are frequently unproductive, except in limestone areas. Occasionally exposures are found, but they are usually deeply weathered and reveal little structural information. Unless an important boundary or other structural feature has to be traced across a ridge, the work of cutting up through the steep forested slopes between the river valleys is in general most unrewarding. Many ridge traverses were however made along well-worn footpaths. In the more rugged mountain areas, where landsliding has breached the subsoil, the underlying weathered bedrock may be extensively gullied. Such places are often difficult of access.

In critical areas, augering was undertaken on the interfluves in order to determine whether a thin weathered sedimentary capping remained; in areas where the rivers have formed their stable profiles in the igneous basement, having cut down through a nearly horizontal sedimentary capping, there may be little indication of the existence of the overlying sediments apart from the occurrence of weathered boulders in the stream beds. For example, Allum (1967) found from a traverse in the Mavo River, western Guadalcanal, that the exposures in the main river were largely of basic igneous rocks, whereas the aerial photographs showed karstic features indicating that limestone outcropped over the greater part of the area. In cases where field observations are incompatible with the preliminary photogeological interpretation, this kind of situation may be responsible.

General observations on Guadalcanal

Communications The administrative capital, Honiara, is situated on the north coast where there is a sheltered harbour on both sides of a small promontory. A motor road extends east and west of Honiara, parallel to the north coast for about 50 km on either side. Three small 'feeder' roads extend from the plains southwards for 7 to 10 km to reach the northern margin of the foothills belt. Also, about 55 km of tractor road extend from Paruru, at the eastern end of the island, westwards along the coast as far as Avuavu.

Air communications have improved considerably since 1964 and now there are airfields at Paruru and Haimarao in south-east Guadalcanal. Small government ships ply around the coast, and on the average there is one official clockwise circuminsular trip every week. Loading from dinghies, particularly on the steep-to south coast, can be hazardous, especially when the south-east trades are blowing strongly.

The larger river valleys of Guadalcanal are reasonable traverse routes, and except in the west-central area, there

Plate 1 View looking north-east from the summit of Sambahatangi (2240 m).
The bevelled summit of Mount Tatuve (1480 m) is on the right, and on the left, on the west side of the Sutakama Valley, is the south-facing escarpment of the Toni Formation

exists a good network of well-established foot tracks which cross the main watershed in several places. The courses of many rivers are interrupted by dangerous rapids and impassable waterfalls, so that time can often be saved by hiring guides who have reliable local knowledge of necessary detours and possible short cuts. In bad weather movement in the headwaters is often delayed for days at a time owing to the continual hazards of flooded rivers, falling trees and landslides. Camp sites have to be selected carefully, preferably on river terraces above the flood level, or near villages, away from the dangers of 'flash floods'.

Geographical names   There has been much confusion in the literature over the accepted spellings of local names: thus the Mbonehe River appears as Bonegi, Bonege and Boneghe. With approval of the Geographical Names Committee of the Solomon Islands Government, efforts have been made to standardise the spelling systems on a phonetic basis (Hackman, 1968a), and it is hoped that this will eventually allay the confusion. In this work the International Phonetic Scheme has been used throughout. Unfortunately this affects some of the stratigraphical names, for example, Tangareso Beds replaces the earlier form, Tangaraisu Beds. Accordingly, where an earlier spelling had become generally accepted in the literature, it follows the standardised form in brackets, e.g. Kaoka (Kau Kau) Bay.

The same river is often given a different name in different parts of its course, according to changes in local dialect. This sometimes leads to confusion: for example, the Vaghanambo, Valasi, Kolohaisava and Lughumboko rivers of eastern Guadalcanal become, respectively, the Kombito, Mbulo, Tavangaoa and Simiu in their lower courses across the plain.

Population   The population of Guadalcanal is about 35 000, of whom over 11 000 live in Honiara. The density of population, particularly in the south, is much greater along the coast than in the interior. The western third of the interior is now uninhabited.

The indigenous people are Melanesians, with dark brown skins, very dark brown in parts of the south-west coast.

Diseases   In general it is safe to drink the clean running water in the mountains and foothills, but the danger of dysentery remains from the stagnant pools and the rivers in the plains. The incidence of malaria has been greatly reduced in recent years by the activities of the United Nations World Health Organisation, and remains a threat only in certain coastal areas. Hookworm is endemic, however, and there is a continual risk of tropical ulcers and boils developing from untended scratches, cuts and poison plant abrasions. Leprosy is no longer a major problem, although it lingers in some of the remoter bush areas.

Agriculture   It is not possible for the average European traveller to 'live off the land' in the Solomons bush. The bush people depend largely on root crops—yams and taro, supplementing their diet occasionally with protein from freshwater fish, possum, wild pig or, on special occasions, domestic pig. The coconut, *Cocos nucifera*, is cultivated in the villages all around the coast, and in larger plantations run

by expatriates; it is the basis of the copra industry. Some cocoa is grown as a cash crop, and even coffee has been cultivated successfully near Komukama in the foothills around the Mberande River.

The cultivation is of the 'shifting' or *ladang* type, the hillsides being burnt off, the major trees felled and the ground left to lie fallow for some months before being hoed with digging sticks. After two or three years the soil becomes impoverished and it becomes necessary to clear new garden areas, and perhaps even shift the whole village several miles away.

On the Guadalcanal Plains experiments are going forward on the cultivation of oil palm.

Vegetation   The forests of the Solomons have been described in considerable detail by Whitmore (1966). The flora is poor in comparison with other areas of tropical rain forest.

The following major types of vegetation can be recognised:

1   *Lowland primary tropical rain forest*, characterised by big-leaved buttressed trees up to 45 m in height, with tall woody climbers. This grades into the next category with increasing elevation.
2   *Mossy forest*, with trees 6–12 m high, with much smaller leaves, and characteristically draped with ferns, mosses and hepatics. The true moss forest tends to be confined to the highest and most exposed parts of the mountain ranges, but grows in the Solomons at much lower elevations than in Papua–New Guinea, where it is not generally found below 2000 m.
3   *Areas of secondary growth*, where the primeval forest has been cleared; these are often characterised by a thick tangle of low-level scrub and thicket which can be very difficult to penetrate. The primary growth has been destroyed up to the summits of some of the higher mountains, e.g. Chaunapaho, 1350 m.
4   *Open heath*, with ferns and bushes, particularly characteristic of the rounded lateritised ridges founded on ultrabasic rocks. There are fewer tree species, the most characteristic being *Casuarina papuana*, *Xanthostemon sp.* and *Myrtella beccarii*.
5   *Grass-covered areas*, notably the western part of the Guadalcanal Plains and the dry terraces behind Honiara, in the lee of the central mountains. Often the boundaries between grass and forested areas are very sharp and unrelated to the underlying rock type. Frequently the grass on the Plains is burned off by the native people, although if fire were excluded, the area might be reafforested. According to Whitmore (1969) most of these unforested areas are probably anthropogenous, maintained by burning after long dry spells.

Fauna   Crocodiles occur, but they are not generally to be considered a danger except in the lower muddy courses of the big rivers, and after dark. They do, however, breed in the middle reaches of the Lungga River. Sharks may be a hazard, even for a few kilometres up the major rivers. One variety of snake is venomous, but the vast majority are harmless. Wild pigs, possums and wild dogs are the largest mammals to be met with in the jungle. Freshwater fish, especially a large species of eel, are a valuable source of protein for the bush people.

Bird life is not as prolific as in most tropical forests—hornbills and eagles are commonly seen in the mountains: wild duck and heron along the rivers.

Insects are abundant; of the other arthropods the most obnoxious is the centipede *Scolopendra metuenda*, which is particularly common in areas with red lateritic soils; it can give an extremely painful bite, which should be treated without delay. A small variety of scorpion also occurs, but is not a common hazard.

Climate   The Solomons enjoy an oceanic tropical climate with temperatures not diverging very much from the range 22° to 32°C. However, in the higher mountain valleys of central Guadalcanal, temperatures of 10°C and even lower may be recorded in the morning hours of a clear starry night. The mean monthly temperature is 28°C and the humidity can be very high, especially in the period of the 'Komburu' between December and March. The average humidity is about 80 per cent.

The rainfall varies from 635 mm per annum on the southern or 'weather' coast to 114 mm on the Guadalcanal Plains; the months of the 'Ara' from July to September tend to be very wet in the south and east of the island, whereas the north and west sides receive more rain at the time of the 'Komburu' between December and March. The most favourable months for field work, particularly as far as the south coast and central mountains are concerned, have been found to be around the changes of the seasons, that is from April to June and between October and late December.

The Solomon Islands have generally been considered to lie outside the cyclone belt; however, during 1967 and 1968 hurricane-force winds were experienced on five occasions in the Solomons region and a considerable amount of damage was sustained.

Geomorphology

Physiographic zones   The physiography of Guadalcanal may be understood in terms of three broad types of regions or landform-systems:
1   The mountain zones.
2   Intermediate or foothill zones.
3   Alluvial zones.

In general the *mountain zones* are characterised by a pattern of very close dissection with deep, narrow ravines, steep slopes and razor-back ridges, although more rounded profiles are characteristic of the ultrabasic rocks and the southern half of the Mbirao Mountains, which are founded on a pile of basic lavas.

The mountain zones occupy most of the southern half of the island (Figure 5). The entire belt might be considered as a deeply dissected composite block extending from Cape Beaufort in the west to Marau Sound in the east. The highest summits are in the central zone, in the Kavo–Haiacha Ranges, being separated from the Suta–Horohana Ranges by the prominent Choruchoru Gap, which follows the line of the Kuma Valley into the Upper Sutakama. Traditionally Popomanaseu, at 2330 m the loftiest part of the long flat-topped Kavo Range, has been regarded as the highest peak in the Solomons. However, according to the contoured AMS maps it would appear that Mt Makarakomburu, the highest peak of the Pukarauvaghalo mass at 2447 m is higher than Popomanaseu. Nevertheless, the more recent DOS 1:50 000 scale maps show that the heights of the two summits are roughly comparable.

The highest mountains form a rugged horseshoe around the head of the Koloula Valley, rising 2000 m over a distance of only 4 km on the south coast side.

Although the edges of this massif are very steep and deeply dissected, the main Kavo Range trending in a northwesterly direction between Mounts Tambunanguu and

Plate 2   View looking eastwards down the Sutakiki Valley. The rounded forms of the Suta Ultrabasics appear on the far side of the Sutakama Valley, in the middle distance. These are overshadowed by the Horohana Mountains, which extend from the bevelled summit of Mount Tatuve on the left to Mount Latinarau (1932 m) on the right

Tanareirei presents an even plateau-like skyline to the north; in fact there is much open level 'moss moor' on the summit areas.

To the east of the Choruchoru Gap are the north-easterly-trending Suta–Horohana Ranges, terminating in the remarkable isolated stack of Mt Tatuve, 1480 m. This mountain, distinctive for its exceptional steepness, is bevelled by a sedimentary capping or surface of marine erosion, dipping steadily at about 25° to the north-north-east (Plate 2).

Tinggetingge (1100 m) is another steep-sided, stack-shaped mountain in the Horohana subzone.

The profiles of the Horohana Mountains are less rugged than those of the Kavo–Haiacha group, being composed of a dissected block of metabasic rocks in contrast to the tough andesites and indurated pyroclastics of the summits surrounding the Koloula Valley. The Horohana Mountains are structurally continuous with the Mbirao zone, from which they are separated by a 600 m saddle between the heads of the Mbokokimbo and Talise rivers. In the Mbirao Mountains are three main summit ridges, each trending in a north-westerly direction *en échelon*; from west to east they are Mt Toghasa (1700 m), Kaichui (1700 m), and the Pinau–Na Suha Range (1625 m). Of these, the last has the appearance of a steeply dissected plateau when viewed from the south or north; Kaichui and Toghasa may be simply the steepened remnants of a more elevated westerly continuation of the same platform. The summit heights decrease towards the east until the range breaks down into the Marau archipelago, a group of islands separated by drowned river valleys and major fault features. Thompson (1960) remarked on the probability of the major channel between Marapa and Tawa'ihi islands being a drowned fault feature which had displaced the ultrabasic outcrop to the north-west.

The northern boundary of the Mbirao Mountains might be extended to include the south-facing escarpment of the Lower Miocene limestone, which rises to 1417 m in Vatupochau on the east side of Lee's Lake. This escarpment forms the northern boundary of a steeply dissected but relatively more 'negative' feature, 1–4 km in width, which can be traced from Lee's Lake eastwards as far as Marau Sound, cutting across the headwaters of the major north-flowing rivers. This is the Mbirao Valley Zone, a belt of relatively less resistant metamorphosed basic rocks.

The western end of the mountain belt, in the Poleo area and lower Itina Basin, is founded on less resistant sediments, although still deeply dissected. In the Marau, Suta and Ghari areas are large ultrabasic masses giving rise to more rounded profiles with the development of fern and *Casuarina* type of vegetation, and a relatively lower drainage density. The drainage density is also lower along the northern fringe of the central–eastern mountain block, and in parts of the adjacent northern foothills zone, in areas of calcareous sediment.

There is a great contrast in the landforms and degree of dissection between the southern mountain zone, founded on pre-Miocene metavolcanics, and the escarpment of Lower Miocene limestone which forms the southern fringe of the foothills zone.

The *intermediate zones* comprise a belt of dissected plateaux running from the north-west corner of the island westward through the centre to form a belt of foothills which narrows

to a width of a few kilometres to the south of Kaoka Bay. In the north-west the Gallego Volcanic Zone is conspicuous as a series of prominent dissected cones rising from an ill-defined platform 200–400 m high. The cone structure is best preserved in the volcanic centre of Mts Esperance and Roundhead, rising to 593 m above sea level. Mts Gallego (1068 m), Paru–Popori, Tanjili and Cone Peak are dissected relics of older volcanoes.

To the south and south-east stretches the monotonous area of the Lungga Plateau drained largely by the headwaters of the Lungga River; the plateau surface slopes north-eastwards from an elevation of nearly 700 m on the watershed to 200 m south of Honiara where it breaks down into an impressive series of elevated terraces, sculptured into the Honiara Beds, a complex of Pleistocene coralgal reef and volcaniclastic sediments.

Broken by the gorge of the Lungga River, this plateau zone may be considered to merge eastwards into the Mbetilonga Basin, a dissected plateau 600–900 m above sea level.

The Mbetilonga Basin is continuous on the east side with the Foothills Zone, 10 to 12 km in width and forming a region of transition between the axial mountain zone and the northern alluvial plains. Transected by the middle courses of numerous large rivers flowing from the mountains, it has a progressive northward slope from about 1000 m to 100 m above sea level, on the southern edge of the plains. This slope is in part a reflection of the dip of the sedimentary pile which is the foundation of the foothills belt. It is possible to trace, in detail, a series of poorly-defined sinuous cuesta scarps, all facing the south.

High-point strip profiles have been constructed from the 1:50 000 series of contoured AMS maps, according to the method outlined by Robinson (1968): the highest elevations within a strip 2 km wide are plotted to give profiles in which the effects of spurious irregularities due to erosion are much reduced. This technique serves to highlight the pattern of late Tertiary–Quaternary block faulting. Figure 6 shows a distinct change in the angle of slope, dividing the Intermediate Zone into two belts, the upper and lower foothills respectively; the angle of slope is steeper and the surface more deeply dissected in the upper belt than in the lower. However, the surface of the upper foothills belt in the central area is bevelled by relics of platforms of erosion, notably on Mt Ndatovitu, the Haviha Ridge and around Kolochulu village.

The northern *alluvial zones* are at their widest between Honiara and Aola Bay—the area known as the Guadalcanal Plains. Essentially the plains area has been built up from the progressive coalescence of the main river deltas which are still actively expanding. The rivers bifurcate and meander extensively in their own alluvium, which ranges in grade from clay to coarse gravel; much of the area is poorly drained. There are large tracts of tall, coarse, rank grass, particularly in the western half of the Guadalcanal Plains.

On the southern or 'weather coast' of Guadalcanal, a narrow alluvial plain extends westwards from Paruru. It becomes progressively narrower west of Avuavu, and west of the Talise River it consists of little more than a shingle beach 50 m wide; its continuity is broken behind Cape Henslow, the southernmost extremity of Guadalcanal, by the hill Keli Ngali, which, although only about 200 m above sea level, stands out as a small but exceptionally rugged, finely dissected land unit, in contrast to the relatively more rounded profiles of the much higher mountains to the north.

Coastal features   Along the western and south-western coasts a series of rugged promontories, Cape Hunter, Cape Austen and Cape Beaufort, jut out into the sea giving a 'cape-and-bay' effect. It seems probable that this part of the coast has suffered fairly recent submergence, and that the drowned valleys have since filled with alluvium; quite extensive tracts of lowland occur at the mouth of the Itina River, also along the lower courses of the major rivers north of Cape Beaufort.

Mangrove and sago swamps extend a short way inland from the mouths of some of the larger west-coast rivers, and there are narrow areas of mangrove fringing some of the island shores in the Marau Sound area, but, in general, Guadalcanal is much freer of coastal swamp than are the other major island groups such as Santa Isabel and New Georgia.

Along the western section of the weather coast, between the Koloula and Itina Rivers, erosion is very active; every stage can be seen in the development of seastacks, for example Vatukulau off the Malagheti coast. Extensive landslides occur in the volcanic cliffs west of Nduindui, while the lower valleys of larger rivers, like the Koloula, are wide boulder-fields made up of colluvial material which has been transported from the upper slopes and continually reworked to form a series of river terraces.

The south coast, locally known as the 'weather coast' of Guadalcanal, is notorious for its rugged aspect; the long fetch across the Coral Sea, particularly in the season of the south-east trade winds, builds up an impressive surf which renders landing from dinghies hazardous. In the local language, this is the Tasi Mauri or 'lively sea', contrasted with the Tasi Mate or 'dead sea' on the sheltered northern coast of the island.

A predominantly westward coastal drift may be deduced from the coastal forms between Lauvi Point and the Aliva-

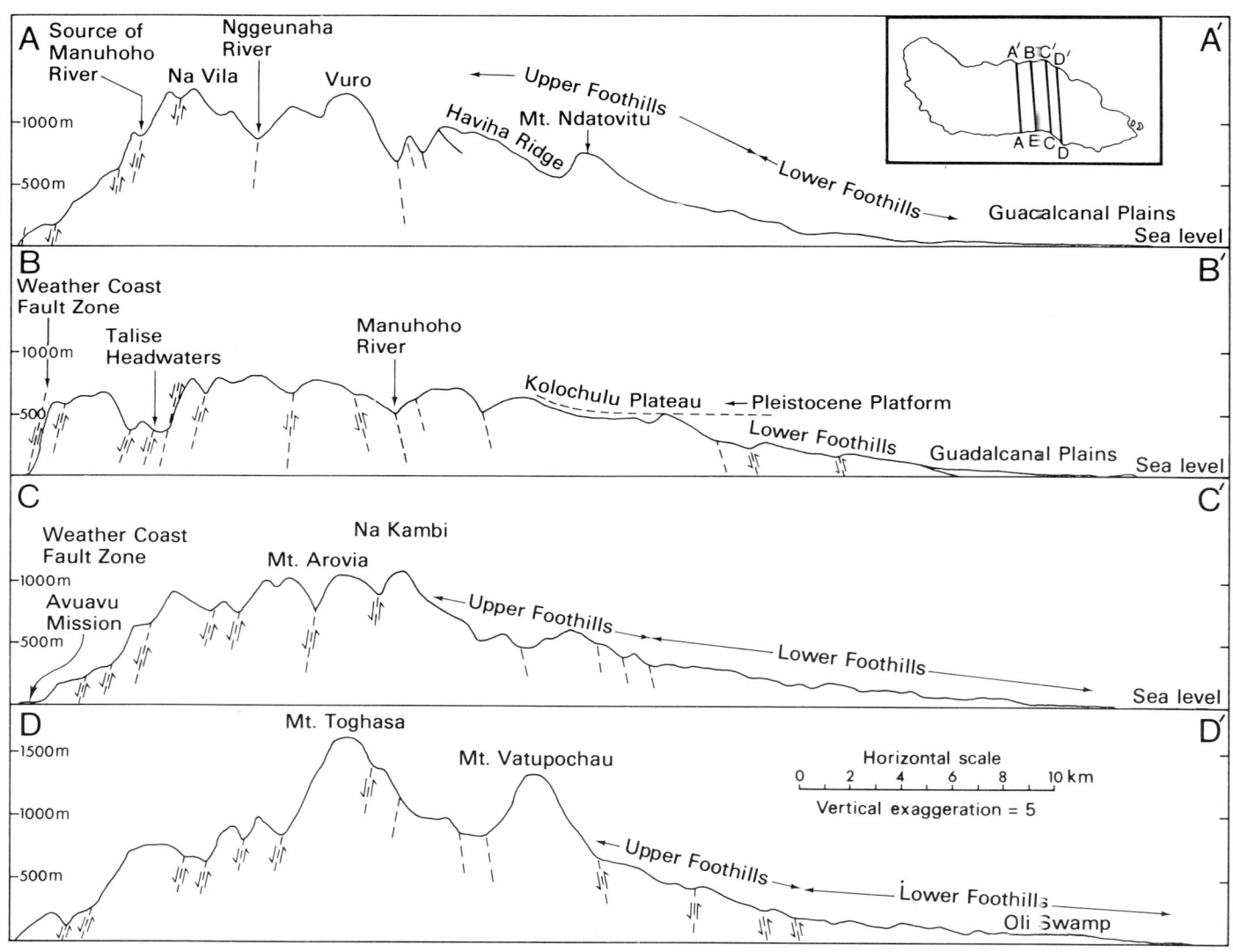

Figure 6   Topographic cross-sections of Guadalcanal

at an elevation of 900 m on the very lip of the watershed: it is separated from the steep upper slopes of the Riva Valley only by a ledge 1 m wide of yellow weathered lateritic clay. The longitudinal profile of the Mbokokimbo–Manuhoho River also shows, in comparison with other major rivers such as the Sutakama, that the steep 'head' of the profile is lacking: in the high-point strip profile (Figure 6, AA') the upper Manuhoho is perched in the 'notch' of a normal fault a considerable distance to the south of the topographic axis of the island; these relationships suggest that the Manuhoho has been beheaded relatively recently by the effect of normal faulting with a large downthrow to the south—here the topographic axis of Guadalcanal is being shifted to the north, while the present watershed lags behind.

A spectacular case of river piracy is seen on the weather coast, where the headwaters of the Kolokoo River have been captured by the Viso. The AMS contoured map in fact shows the head of the Viso as flowing into the Kolokoo, but this is an error. The Kolokoo is a grossly underfit stream, braiding in a wide boulder-strewn valley with a wind gap at its head, whilst the Viso appears to be 'overfit', a large volume of water passing through a series of narrow rapids and deep ravines between the 800 and 300 m contour levels.

Most of the north-flowing rivers show nickpoints for which the projected base levels are at approximately 45 and 140 m above sea-level; these correspond to two of the most conspicuous levels in the sequence of terraces behind Honiara. The river profiles in south-east Guadalcanal reflect the sequence of step-faults parallel to the weather coast.

The north-flowing rivers meander where they cross the plains in their lower courses; ingrown meanders (Thornbury, 1954, p. 145) occur in the middle or 'foothill' reaches, for example, in the Sutakama and Mberande rivers. These rivers, on leaving the pre-Miocene 'basement' zone cut deep consequent gorges through the Miocene limestones at the base of the sedimentary succession. These gorges appear to act as 'traps' for the boulder load, so that in the lower courses the proportions of different floaters often give quite a misleading impression of what sort of geology to expect in the headwaters.

Ingrown meanders are prominent in some of the rivers above the limestone gorges, notably in the Kombito, Rere and Nggeunaha rivers. It might be misleading to construe this as an effect of local uplift; apparent greater maturity could be an expression of the relatively greater erodibility of the underlying rocks, the metavolcanics of the Mbirao Valley Zone.

Meanders are also well developed in the middle and lower reaches of the Lungga; a spectacular gorge up to 200 m deep has been cut in the Miocene–Pliocene succession south of Mt Austen. During investigations for a proposed dam site, drilling has shown the rock floor of the gorge to be over 50 m below the present river bed. This suggests that the Lungga has recently recovered from a phase of overdeepening which may have been connected with a fall in sea-level during the Pleistocene.

Throughout the Mountain Zone, the minor tributary valleys are frequently discordant, with spectacular waterfalls developed above the junctions with the main rivers. Deeply incised V-shaped valleys, with interlocking spurs, are typical.

The lower courses of the bigger rivers are subject to sudden alterations; in 1954 Grover noted that the Rere River had been captured by the Susu on the plains inland from Rere Point, leaving the Rerembonumbonu as a minor divergent branch (Grover and Pudsey-Dawson, 1958). In 1957, an exceptionally high flood breached the banks of the Susu below Sani Lumu, and the lower Rere established a new course: it now disembogues 2 km west of Rere Point.

Landslides and lakes   Landslides and downward vertical erosion are the most important sculpturing agents in the Mountain Zone. Landslides are particularly common on the flanks of the Haiacha Ranges and in the upper part of the Koloula Valley. Extensive debris slides of fresh angular plutonic and volcanic rocks flank the tributary valleys of the Koloula and feed the angular boulder fields of the main valley.

In the upper reaches the landslide profiles are steep, but by gradual transport of the material downstream, and under the influence of gravity, the colluvial material spreads out to form a series of well-defined terraces, and is in part reworked by the main river.

In July and August 1965, several days of exceptionally heavy rain, combined with a period of increased seismic activity, triggered off a series of landslides on the weather coast and in the upper Sutakama Valley. A spectacular slide occurred at the head of the valley of the Charikimberi tributary. This was a rock slide in the sense of Sharpe (1938) involving the dislocation and sliding of an individual rock mass down a series of joint surfaces. The writer had the opportunity to observe the upper Sutakama Valley in May/June 1965, and again in December of the same year; before the rock slide the upper Sutakama River had been very difficult to negotiate on foot, the river bed being choked with large slippery boulders of igneous rocks, separated by deep pools and rapids. The rock slide occurred in metavolcanics with well-defined planes of schistosity dipping at angles of 50° to 80° to the north-west. Thousands of cubic metres of material slid down the Charikimberi to build a colluvial fan at its confluence with the Sutakama. The valley floors are extremely narrow here, and pronounced headward erosion had undercut the steep lower slopes of the Latinarau Range.

This material had blocked the main river, ponding up the water behind. Over a period of three months 20 cm of fine blue silt and mud had been deposited in the middle of the temporary lake. By December the river had transported a good deal of the finer material downstream as far as Nuhu, forming a smooth terrace of up to 10 m in height, the grain size progressively decreasing downstream. Walking along the river bed was in consequence considerably easier than it had been six months before. Below Nuhu, the muddy material continued to be winnowed out, and the lower Mbalisuna River was sullied with sulphurous mud (derived from pyritised metavolcanics) for a further two years as a direct consequence of this rock fall, even during prolonged dry periods.

In Guadalcanal landslides are also common along the escarpment of Lower Miocene limestone or calcarenite, where it overlies the northern side of the pre-Miocene outcrop. Ordinary rockfalls occur in bad weather in the limestone gorges, or may be triggered off by earthquakes. At the end of the last century, there was a landslide at the head of

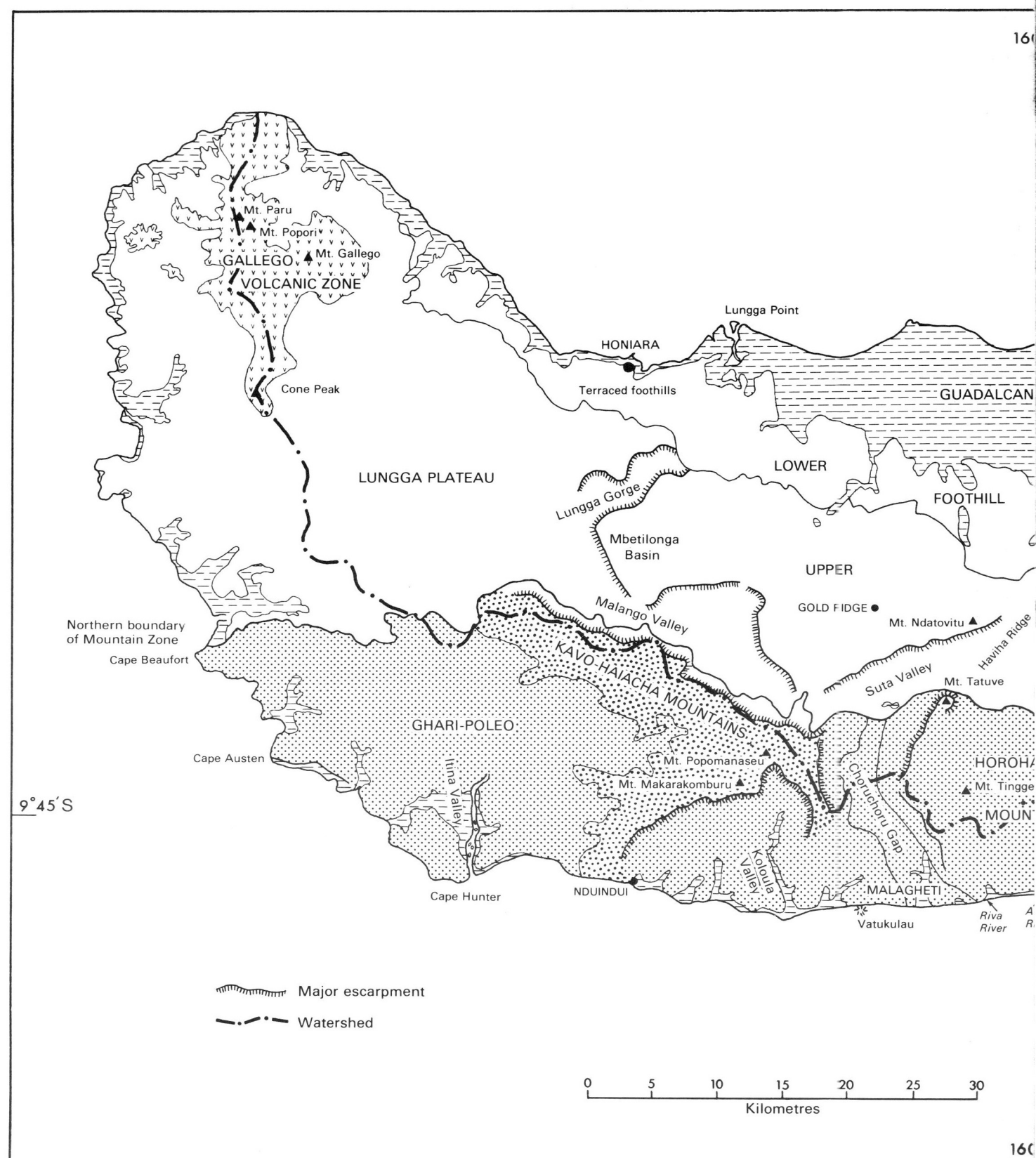

Figure 5   Map of physiographic zones in Guadalcanal

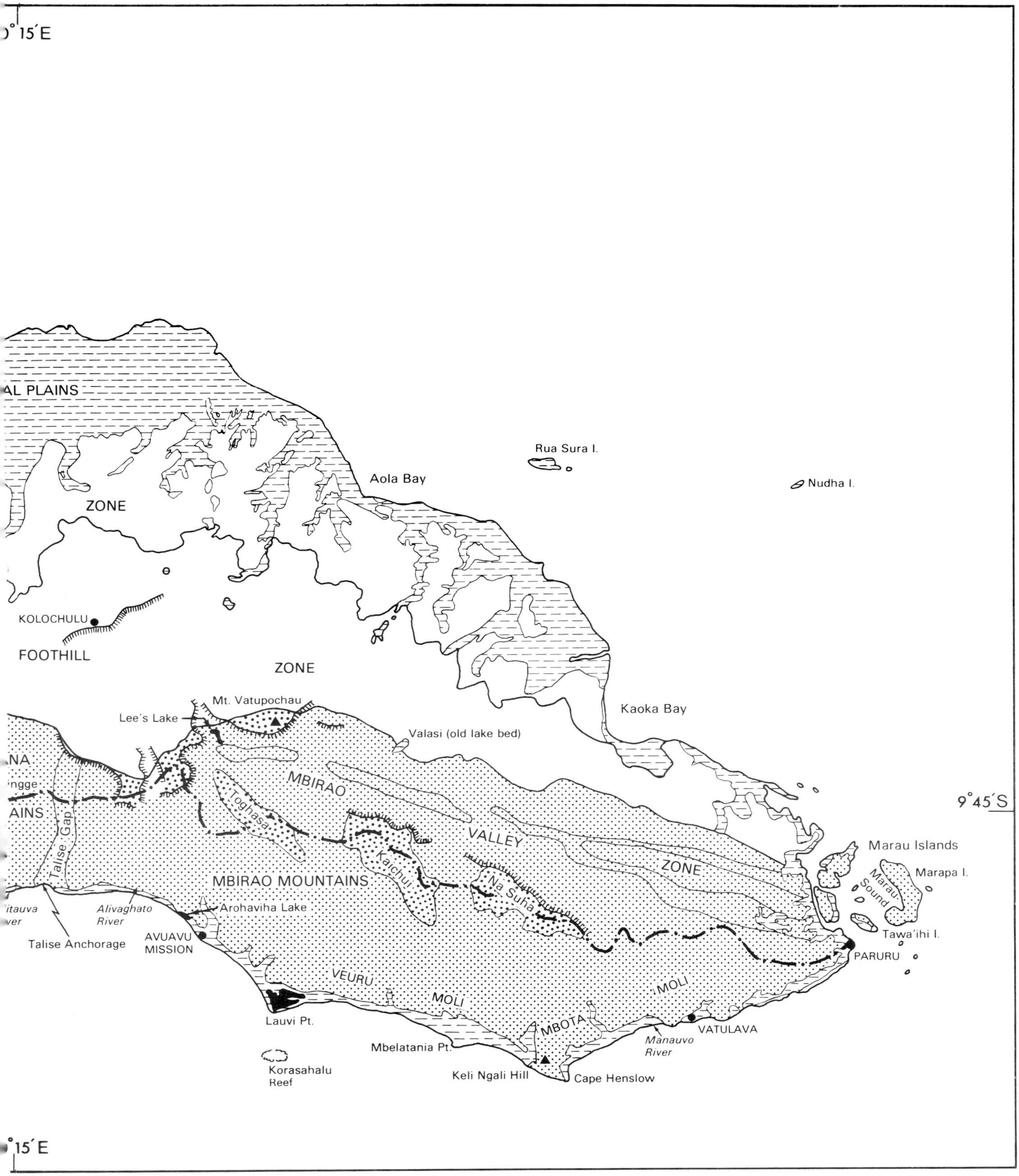

AL PLAINS

ZONE

Aola Bay

Rua Sura I.

Nudha I.

KOLOCHULU

FOOTHILL

ZONE

Mt. Vatupochau

Lee's Lake

Valasi (old lake bed)

Kaoka Bay

NA

ngge

MBIRAO

9°45′S

AINS

Talise Gap

Toghasa

VALLEY

ZONE

Marau Islands

Marapa I.

MBIRAO MOUNTAINS

Kaichui

Na Suha

Marau Sound

itauva
ver

Alivaghato
River

Arohaviha Lake

Tawa'ihi I.

Talise Anchorage

AVUAVU
MISSION

PARURU

VEURU

MOLI

MOLI

Lauvi Pt.

MBOTA

Manauvo
River

VATULAVA

Mbelatania Pt.

Korasahalu
Reef

Keli Ngali Hill

Cape Henslow

ghato River. The mouth of the Haimatua River has been diverted for 3 km to the westward along the line of the Kolotambu swamp, barred from the ocean by a shingle spit 100 m wide. Similarly, to the west of Avuavu a small lagoon, the Arohaviha Lake, has been cut off from the sea by a westward-growing spit. Assuming that the direction of prevailing wave drift corresponds with the direction of maximum fetch, that is from the south-east, this would appear also to coincide with the observed direction of longshore current.

Lauvi Lagoon is bounded on either side by a series of at least three spits which have nearly coalesced to form an elaborate cuspate point. The spits on the western side are longer and built of coarser shingle than those on the eastern flank. It is suggested that the reason for the development of this point is in part the sudden change of exposure to the direction of advance of the dominant storm waves. Longshore drift from the east, combined with wave action from the south-east, first built up a 'hooked' spit which eventually reunited with the coast further west. The occurrence 3 km offshore of the Korasahalu Reef may also have been influential in acting as a partial barrier to the main waves from the south, so that a dominant south-easterly wave drift operated on the east side, combined with a subsidiary south-westerly drift on the west side. Mbelatania Point, 14 km to the east may be a similar case, but on a smaller scale.

Recent coral reef does not occur on the north coast of Guadalcanal between Lungga Point and Aola Bay, where the big rivers crossing the plains raise the level of turbidity in the offshore zone. No major rivers sully the recently submerged area of Marau Sound, however, where coral reefs and clear water fringe all the islands. Patches of coral grow around the indented coastline of north-west Guadalcanal, mostly in the shelter of small bays and offshore rocky reefs. Coral has grown on the weather coast only on the sheltered eastern side of Talise Anchorage, and in comparable situations further east. It has grown more steadily along the stretch between Cape Henslow and Paruru, a part of the coast which faces the western end of San Cristobal, 80 km distant, and which therefore does not receive the full force of the south-easterly fetch across the Coral Sea.

Stoddart (1969a) has described the geomorphology of the coral reefs of Marau Sound: a remarkable feature, still not satisfactorily explained, is that the contemporary corals on the modern reef flats are invariably dead.

Rua Sura, Sura Kiki and Nudha islands are essentially cays of broken coral sand overlying more coherent recrystallised coralline limestone. The islands lie on an east–west 'swell' running from 6 to 20 km off the north coast of Guadalcanal.

In the Mbota Moli area, the highlands sweep right down to the sea, interrupting the alluvial strip between the Manauvo and Vatulava rivers, and also near Oa.

The patches of coralline limestone fringing the coast between Cape Henslow and Paruru represent a Recent offshore reef which has been elevated about 10 m: the lagoon behind has been filled with gravel and silt brought down by the mountain torrents. It seems that much of the weather-coast plain is a ledge that has been formed in this way, although near Sukiki the reef is obscured by shingle accumulating from longshore drift, and westwards from this point

there is very little development of coastal reef limestone apparent, except in Talise Anchorage.

In summary, the evidence from the outline of Guadalcanal shows that apart from the eastern (Marau Sound) area of recent submergence, the coastline appears to be presently 'neutral', that is predominantly neither emergent nor submergent.

Drainage There is a very dense network of drainage on Guadalcanal; seven major rivers drain the area north of the watershed to meet the north coast between Honiara and Aola Bay. Of these, the combined Mbokokimbo–Manuhoho system drains the largest area (about 365 km²), and is also the longest river in the Solomon Islands (92 km). The Manuhoho rises only 1.6 km from the south coast, the watershed being much closer to the weather coast than to the north coast, especially in the eastern and central sectors. None of the rivers of Guadalcanal could be described as navigable; on the southern side of the watershed their upper courses are torrential in aspect, and the lower valleys are often more or less dry except in time of flood, the valley floors being strewn with boulders derived from transported colluvial material.

The main rivers flowing down to the Guadalcanal Plains have a distinct tendency towards parallelism along north-easterly lines, parts of their courses being so straight that control by fracturing is clearly indicated. A north-westerly trend is also evident in some river courses, for example, the middle reaches of the Manuhoho and the upper sections of the Hanangga and Sambahalava rivers.

Faulting on west-north-westerly lines has clearly deflected the middle courses of the Sambaharihi and Mbolavu rivers on the south coast. In particular, the 'shutter ridge' effect (Cotton, 1958, p. 167) on the upstream side suggests that there has been transcurrent movement along this fault line.

The general drainage pattern, particularly in the northern foothills area, is of the 'open trellis' type. However, in the less resistant parts of the mountain belt, along the line of the Mbirao Valley system, there is a tendency for the drainage to fan out into a more dendritic arrangement as, for example, in the headwaters of the Simiu and Kolohaisava rivers. The area of the Lungga Plateau drained mainly by the Lungga, Poha, Mbonehe and Mavo, has a well-defined angulate pattern, which contrasts with a distinct radial tendency around the Gallego volcanic centres in the north-west of the island.

The upper course of the Nggeunaha–Mberande River, in south-central Guadalcanal, exhibits a 'barbed' pattern, suggesting that the headwaters may have gained at the expense of the rivers flowing to the south coast. However, capture in the reverse direction is suggested by relationships in the upper Manuhoho Valley; headward erosion by the Riva appears to have captured the source waters already, and the Alitauva is also likely to behead the Manuhoho still further if headward erosion continues actively to the north-west— the steep gradient, with the risk of landslides, is a distinct advantage in favour of the south-coast drainage. The uppermost section of the Manuhoho is a misfit stream, in the sense that its valley is wide and mature in relation to the small flow of water. The present source of the Manuhoho is an almost stagnant pool, the Tangisi Koivo, which is situated the Vuraka tributary of the Mongga. The material choked the upper part of the Mongga gorge, and Lake Lee or Na Kuma, a shallow body of water about 1250 m long and, at the most, 250 m wide, was ponded up behind the obstruction. Drowned trees may still be seen in the middle of the lake.

Areas of calcareous rock   The karst labyrinth, with the 'pustulose' appearance on aerial photographs which is typical of the weathering of raised limestone areas in the wet tropics, and which can be found in other islands of the Solomon group, for example Choiseul, where there are large areas of uplifted Pleistocene limestone, is not found as such in Guadalcanal, although parts of the outcrop of the Honiara Beds and Mbonehe Limestone in the north-west of the island show some tendencies towards that kind of 'cockpit karst' pattern.

In the southern part of the foothills belt, a large proportion of the country is founded on Lower Miocene calcareous rocks. Thin-bedded, well jointed and fissured calcarenites might be expected to give rise to some features of karstic scenery; in fact only on the escarpment of Vatupochau are these developed to any marked degree; there, a pavement of solution fretwork, etched, grooved and riddled with chasms, is obscured by dense undergrowth—an area so tedious and dangerous to traverse that it is religiously avoided by the local inhabitants. Systems of underground drainage occur, but none have been explored in detail; notable are the cave systems of Mbetilonga, where the Tenaru River flows underground for a distance of about 800 m, and the caves of Kolokokua (meaning 'the water that hides itself'), a small tributary of the Kombito. In areas adjoining the outcrops of calcareous formations, the rocks are often obscured by such thick layers of travertine that it is not possible to map the bedrock geology.

In the Mountain Zone, even very narrow outcrops of recrystallised limestone (the Tetekanji Limestones) within the basement volcanics give rise to localised karst features. An example is the doline of Paluluna in the Mandonu headwaters of eastern Guadalcanal, which is a solution hollow 20 m in diameter and about 15 m deep.

The lake of Na Hoeta, at the head of the Na Hue River, is also in part a solution hollow, lined with crags of well-jointed white recrystallised limestone.

TWO
Stratigraphy

The rocks of Guadalcanal comprise two major series—the pre-Miocene basement, essentially consisting of more or less strongly folded and metamorphosed 'oceanic' basalts with intrusive 'alpine'-type ultrabasics, and a Miocene to Pleistocene series, up to 6000 m thick, of unmetamorphosed flat-lying to gently folded volcanic and sedimentary rocks. The sequences in various parts of the island are tabulated in Figure 3 and the lateral facies variations in northern Guadalcanal are illustrated by a schematic cross-section in Figure 4.

THE PRE-MIOCENE BASEMENT

The term 'pre-Miocene Basement' has commonly been used by geologists working on Guadalcanal; it comprises the Basement Complex as described by Coleman (1960b) together with the Ultrabasic Instrusives first mapped in western Guadalcanal by Pudsey-Dawson and Thompson (1958).

The pre-Miocene Basement is here subdivided into:
1   The Mbirao Group (new name).
2   The Guadalcanal Ultrabasics.

THE MBIRAO GROUP

The Mbirao Group is named for the ethnic area over which the greater part of its outcrop extends: essentially it is a thick sequence of basic volcanics, subsidiary limestones and doleritic sills, intruded in depth by augite–labradorite gabbro.

The following subdivisions (all new names) have been mapped:
The Mbirao Metabasics
The Guadalcanal Gabbro
The Mbirao Dolerites
The Tetekanji Limestones
The Mbirao Volcanics

There is sufficient evidence to support the view that the Mbirao Metabasics, in part the equivalent of the Guadalcanal Schists of Thompson (1960), are essentially metamorphosed derivatives of the other subdivisions of the Mbirao Group; thus basaltic volcanics, doleritic and gabbroic intrusives may all be identified within the metabasic belt, but only the Tetekanji Limestones subdivision can be mapped as a distinct unit therein. No absolute evidence is yet available for the age of the Mbirao Volcanics. The Tetekanji Limestones, interbedded with basaltic lavas, are extensively recrystallised and have yielded no recognisable microfossils.

A K:Ar determination on a sample of Guadalcanal Gabbro (No. 65532) from the Vaghanambo River yielded an age of 92 ± 20 million years, that is probably Cenomanian (Snelling and others, 1970).

Tentative comparisons may be made with 'basement' rocks from other parts of the Central Province. Thus an age determination of 66 ± 3 million years for an intrusive basalt on Santa Isabel gives an uppermost Cretaceous to basal Palaeocene date for the hypabyssal phase: also the basement Ravo Limestones of San Cristobal have yielded *Globorotalia spp.* of probable Palaeocene–Eocene age (Coleman, oral communication).

The pre-Miocene amphibolites of Choiseul Island have been dated as lower Middle Eocene (50 million years) from the work of Richards and others (1966).

If analogies with Guadalcanal hold good, the Mbirao Volcanics may have been extruded in the Cretaceous and metamorphosed in the Eocene. Although a long-range, tenuous correlation, this would not conflict with the geological data.

The basic volcanics of south-east Guadalcanal do not appear stratiform on the aerial photographs, and when viewed from the sea many of the summit ridges show relatively rounded profiles; accordingly some observers have been led to suspect the presence of gabbro and ultrabasic rocks rather than basalt (Thompson, 1960). In detail, however, the drainage is coarse and deeply incised with razor-backed interfluves. The photographic tone is fairly dark, in contrast to the light tone of the ultrabasics; the only discernible structures are the major lineaments, which have a profound affect on the drainage pattern over the entire outcrop.

The areas of Tetekanji Limestones are too limited to be recognisable on 1:50 000 aerial photographs. The outcrops of Mbirao Dolerites and Guadalcanal Gabbro cannot be distinguished photogeologically from those of the Mbirao Volcanics, but the average tone of gabbroic areas is somewhat lighter than that of basaltic areas. The metamorphosed belt may be distinguished from the adjacent Mbirao Volcanics by the greater abundance of lineaments and lesser resistance to erosion; however, only along the line of the Mandonu Fault in the east can the boundary be clearly delineated. To the north of this fault numerous fine parallel lineaments impart a distinct west-north-westerly 'grain' to the Mbirao Metabasics outcrop.

The Mbirao Volcanics

The term 'Mbirao Block' is used for the entire basement area east of the line of the Suta Ultrabasics. The Weather Coast and Valasi 'Blocks' may be regarded as subsidiaries thereof, corresponding to areas of relatively unaltered basaltic lavas, and separated by a strip of regionally metamorphosed basic schists. Small areas of unaltered basalt have been mapped in the Na Hue tributary of the Lughumboko River, the Aola headwaters and the Nggeunaha section of the Mberande River. Other outliers could probably be detailed on the higher mountains north of the metabasics boundary, at the expense of much laborious climbing over the rugged summits.

The Alualu and Hanangga rivers provide good sections through the Weather Coast Block, while the Rere and Valasi river sections are typical of the Valasi Block.

The dip of the stratification has been measured in the relatively few exposures where good pillow formation can be inspected. In the Weather Coast Block the lava pile dips generally at 20° to 30° towards the north, but is progressively downfaulted to the south by a series of east–west step faults. Along their southern boundary the Mbirao Volcanics are overlain by the alluvium of the coastal strip, and between Cape Henslow and Oa they are covered by a veneer of Pleistocene sediments, the Mbota Moli Beds. On the northern side there is a gradual transition from unaltered basic lavas to metabasics of the greenschist facies: the Valasi Fault, trending west-north-west, clearly defines the boundary between the Mbirao Metabasics and the relatively unaltered basic lavas of the Valasi Block to the north.

In the Vaghanambo River, Pliocene calcarenite rests unconformably on a surface of basalt, there being no evidence for a long intervening period of subaerial erosion. Unaltered basaltic lavas are exposed over about 800 m of the Nggeuna-

Plate 3 Basaltic pillow lavas with interstitial epidote, quartz and calcite. Mbirao Volcanics, Kolokiki tributary of the Sambahalava River, Weather Coast Block

ha River, 23 km to the west of the Valasi Block; 1.2 km north of the village site of Haimela the amphibolites of the metabasic zone have been thrust to the north over fresh unaltered blue-grey basalts, along a plane trending west-north-west and dipping steeply to the south; this small area of unaltered basic lavas may be regarded as analogous in some respects to the Valasi Block.

In the Weather Coast Block the lava pile is estimated from vertical sections and high-point strip profiles to be at least 1200 m thick. The total thickness of the Mbirao Group, including the metabasics, must be much greater.

Pillow lavas, petrographically clinopyroxene–labradorite basalts, predominate throughout the Mbirao Volcanics (Plate 3). Many of the outcrops appear massive, however, and typically consist of monotonous grey-weathering fine-grained basalt, bluish grey or pale blue on freshly hammered surfaces. Usually jointing is closely spaced in heavily faulted areas, irregular in aspect elsewhere; columnar effects are developed locally in the Oa River. The measured diameter of individual pillows varies between 15 cm and 2 m. In the Valasi Block the basalt is usually purple or reddish due to finely divided haematite, some outcrops showing an attractive green and purple variegation; but pillow formation is not as well developed as in the Weather Coast Block.

Minor bodies of pink and white recrystallised limestones and multicoloured cherts are not mappable on the scale of 1:50 000; epidote, quartz and chlorite commonly fill the interstices of the pillows. Locally the lavas are autobrecciated and intricately veined by calcite and quartz; the assemblage quartz–calcite–chlorite–epidote is commonly encountered

Plate 4 Photomicrograph of porphyroid chert, Mbirao Volcanics. Kolovaghamela River, eastern Guadalcanal. Plane polarised light

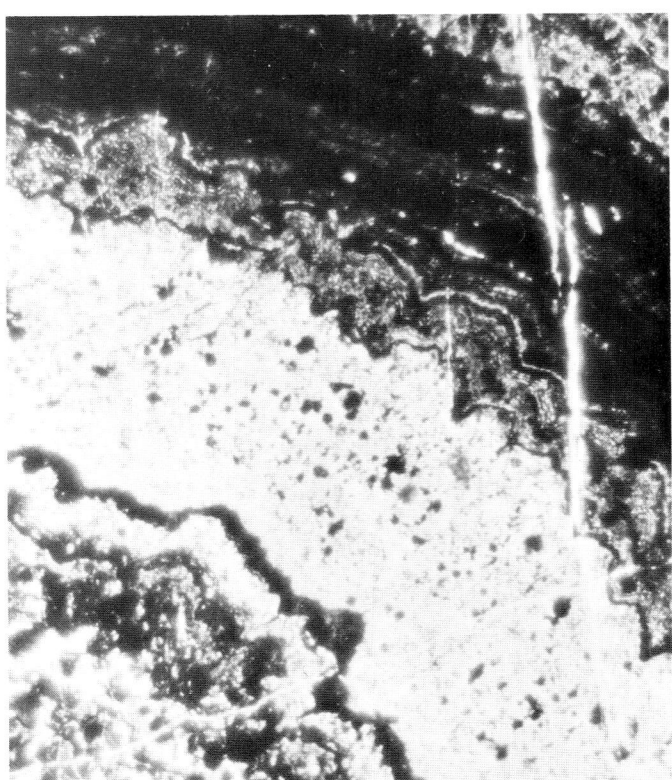

in composite veins. Pyrite, pyrrhotite and actinolitic slip-fibre have developed locally on shear planes.

In thin section a typical pillow lava is a fine-grained or dense basic rock; stubby plagioclase laths up to 0.3 mm in length are usually too altered to allow for optical determination. Augite granules approach 0.2 mm in diameter, but the entire section is obscured by grass-green epidote and the brownish coloration of the glassy matrix.

The texture of the Mbirao basalts is typically intersertal, but may vary to hyalo-ophitic, intergranular or microporphyritic. The clinopyroxene, often altered to chlorite and actinolite, is augite with an average value of 60° for 2V. Saussuritised plagioclase comprises up to 65 per cent of the identifiable constituents, the compositions (as determined from twinning by the Michel-Lévy method) varying from An $_{20}$ to An $_{50}$ (oligoclase–labradorite range).

Subordinate hyaloclastite is composed of turbid basic glass fragments with palagonite reaction rims 10 to 25 mm thick: skeletal pyrite is concentrated in zones peripheral to the shards. Another type of breccia contains abundant vesicles up to 2 cm in diameter set in a highly altered matrix: the cavities are lined with pumpellyite and zeolites.

Locally, in the head of the Kolovaghamela River, there are substantial bodies of attractive red and brown porphyroid or spherulitic cherts.

The Tetekanji Limestones

These limestones, which are mappable as discrete minor bodies probably nowhere exceeding 200 m in thickness, within the Mbirao Group, are named for the Tetekanji subdistrict of eastern Guadalcanal.

Two remarkably linear belts of limestone may be mapped:
1   A northerly belt, extending from near Mboranaho on the upper Rere River for 40 km west-south-west to the northern end of Wahere (Komachu) Island in Marau Sound;
2   A southerly belt, forming a chain of narrow outcrops along the line of the Mandonu Fault.

To the west, minor occurrences of limestone interbedded with schist have been recorded from the middle Nggeunaha River and the Charikimberi tributary of the upper Sutakama.

A section in the Na Puleo tributary of the Kolovaghamela River, probably the one first visited by Grover and Pudsey-Dawson (1958), which reveals 20 m of massive white recrystallised limestone, dipping at about 25° to the north-west, is taken as the reference locality. Upstream is a broken succession of about 30 m of red and pink laminated limestone, transected by thin veins of white secondary calcite.

In the Na Puleo section the limestones appear to be interbedded with lavas and metabasics of the Mbirao Group in normal sequence, although elsewhere major faulting has frequently been observed to coincide with the margins of the limestone outcrop: for example, along the line of the Mandonu Fault.

The Tetekanji Limestones are typically white or pink sheared recrystallised carbonate rocks. Occasional remnants of planktonic foraminifera are recognisable, but to date could not be specifically identified.

Stratification may be observed where there are tabular, broadly concordant layers of chert and thin basalt flows within the limestones. Often discrete layers of pink and white sheared limestone alternate as tectonic slices with sheared and brecciated metabasalts. Occasional thin interbeds of dense grey pelite trace out intricate drag folding effects in the metabasic zones.

In the Na Vuvuti area in the headwaters of the Na Hue River, north-east Guadalcanal, thin beds of brecciated limestone have been stained red and black due to finely divided iron and manganese oxides. Replacement pockets of grey powdery manganese oxides or earthy wad also occur within the limestone.

The absence of coarse clastic material, the identification of remnants of planktonic fossils, and the rarity of pelitic intercalations, suggest that the Tetekanji Limestones were predominantly of pelagic origin, deposited in the bathyal zone, and comparable with the calcareous ooze now covering much of the south-west Pacific floor (Revelle and others, 1955).

As discontinuous chains of outcrops are mappable over a distance of at least 40 km, these limestones probably represent fragments of a formation which accumulated during a temporary cessation of the Mbirao vulcanicity. The depositional environment was also favourable for the segregation of chert and manganese oxides.

The Mbirao Dolerites

Minor bodies of dolerite or microgabbro are intrusive into the lavas and gabbro of the Mbirao Group:

MAJOR SILLS   The largest dolerite bodies outcrop in an east–west belt about 4 km inland from the 'weather' coast, between the Mbolavu and Oa rivers. Boulders found in the headwaters of most of the major rivers show that similar intrusives may remain unmapped at higher levels in the mountains.

From meagre evidence of contacts in the Kolilua River and the parallelism of flow banding with the stratification of the lava pile, the larger bodies are thought to be sills within the Mbirao Volcanics. Unfortunately where the boundaries of the dolerite bodies are exposed, they are usually highly sheared or faulted, so that relationships with the surrounding lavas are obscure.

The largest sill, which is exposed in the Hairughu and Sambaharihi rivers of the Veuru Moli area, is 400 to 500 m thick, with a horizontal extent of 5 km along the strike.

In hand specimen the dolerite is typically a tough green phanerocrystalline rock, with an attractive ophimottled effect. Light green patches 1 to 2 mm in diameter represent pyroxenes which ophitically enclose plagioclase laths: the background is made up of saussuritised plagioclase. Usually the rock is massive or irregularly jointed, although columnar effects are seen locally in the Kolopura River, Mbota Moli. The texture of the dolerite is commonly ophitic, but coarse-grained varieties are subophitic to hypidiomorphic-granular. In some coarse segregations pyroxenes may reach a length of 2 cm. The feldspar is usually turbid and, although twinning indicates a compositional range of An_{44} to An_{60}, refractive index tests indicate alteration to a more sodic plagioclase.

MINOR BASIC DYKES   Linear bodies of dolerite with a north-easterly trend have been mapped within the Weather Coast Block and in the metabasic zone to the north. As the margins are obscured, they cannot be distinguished from fault-bounded exclaves of dolerite sills. However minor discordant bodies of the same trend demonstrably intrude the fresh basic rocks of the Valasi Block, notably in the middle reaches of the Vaghanambo and Rere rivers. The largest of these bodies is not wider than 30 cm. Several larger dyke-like bodies of microgabbro have been mapped in the upper Sutakama Valley; they trend north-east, and can be regarded as cognate with the Guadalcanal Gabbro into which they have been emplaced.

Plate 5 Dolerite dyke with feldspar glomerocrysts showing flow-alignment, in Mbirao Group, Weather Coast Block

GLOMEROPORPHYRITIC FELDSPAR-DOLERITES Seven examples of discordant bodies varying in width from 3 to 20 m have been mapped in southern Guadalcanal between the upper Manuhoho and Sambaharihi rivers. They intrude both the Mbirao Volcanics and the metabasics.

Porphyroid clusters of cloudy white saussuritised feldspar up to 30 cm in diameter are embedded in a speckled greenish matrix of altered dolerite; the boundaries of these glomerocrysts may be crudely polygonal, spherical or ellipsoidal, with the longest axes parallel to the walls of the dyke (Plate 5). The density of the glomerocrysts may vary considerably, even across a single dyke.

In thin section the doleritic matrix is found to be composed of granular areas of epidote, clinozoisite, pale green actinolite, turbid plagioclase, leucoxenised ilmenite and sphene.

Possibly the feldspar glomerocrysts indicate an early phase in the separation of an anorthositic fraction from basaltic magma, in which plagioclase crystallised in radially arranged aggregates.

Plate 6 Photomicrograph of leucogabbro from the Guadalcanal Gabbro, Valasi Block. Crossed nicols. Middle Rere River, east-central Guadalcanal.
Bytownitic plagioclase, showing some evidence of protoclastic deformation, comprises up to 60 per cent of the rock. Pale green uralitised pyroxene is interstitial to the feldspar; iron ore is practically absent

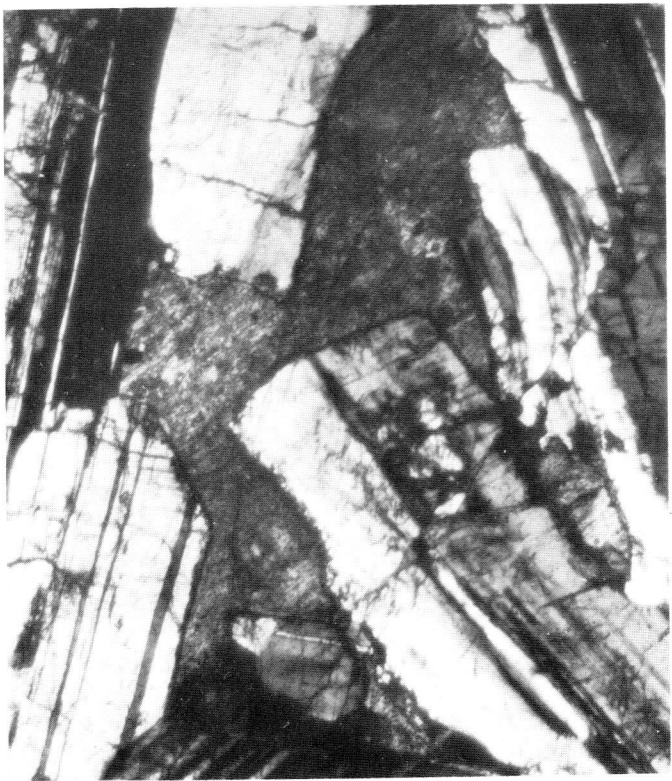

The Guadalcanal Gabbro

Gabbro outcrops flank the basement metabasics on the north in eastern Guadalcanal, and on the north-west side in central Guadalcanal. In both cases the boundary is a major fault zone intruded by ultrabasics. Gabbro also appears in south-west Guadalcanal as a narrow strip about 12 km long flanking the southern side of the Ghausava Ultrabasics (Thompson, 1960). Scattered outcrops of gabbro are mappable within the Mbirao Metabasics. It may be surmised from the outcrop pattern that gabbro forms the greater part of the basement area which is obscured by the Miocene–Recent succession in northern and western Guadalcanal.

The margins of the gabbro are generally highly sheared; however in the Valasi Block the gabbro is demonstrably intrusive into the Mbirao Volcanics, where the relationships are suggestive of emplacement by stoping. In the headwaters of the Mbicho River, the gabbro contains xenoliths of basalt petrographically identical with that of the Mbirao Volcanics.

The complex relationships between Mbirao Volcanics, gabbro and a later hypabyssal phase are well exposed in the Rere River 1.5 km above Isipeni village, and in the Vaghanambo gorge to the west. The dykes are sinuous and often their margins are ill-defined; locally they have been ruptured into subparallel streams of xenoliths or show drag effects, presumably due to mobilisation of the surrounding gabbro. All stages of ingestion of dyke material by the leucogabbro host can be seen. It may be inferred that the host gabbro was still plastic when the dykes were emplaced.

Where undisturbed, post-Oligocene sediments rest nonconformably on the gabbro, although locally, as in the Tina River, the boundary has been tectonised by post-Miocene movements. The Guadalcanal Gabbro is typically a tough, grey weathering, coarse basic rock, with a 'variegated' appearance due to the contrast between the purplish grey pyroxene and the greenish tinge of saussuritised feldspar; polished sections give this 'purple and green' effect in reflected light. Examination of thirty thin sections from different areas has shown only minor petrographic variations. Plagioclase constitutes about 60 per cent of the rock (Plate 6), varying in composition from An_{50} to An_{71}, i.e. the labradorite range: only one thin section showed bytownite (An_{80}).

Clinopyroxene forms 30 to 40 per cent of the rock, but has usually been altered to actinolite and chlorite; patchy relics of original clinopyroxene are of a pleochroic brownish variety, probably titaniferous augite. Common accessories are magnetite, leucoxene, sphene and pyrite. Protoclastic effects, such as marginal granulation of crystals and strained twin lamellae, are particularly noticeable in thin sections of gabbros from the eastern parts of the outcrop.

The main variations in the field aspect of the gabbro are conditioned by the following factors:
1 The colour ratio of 20 to 40 is more leucocratic than that of the average gabbro. Locally, finer-grained, more melanocratic facies are found: in the upper Sutakama gorge these occur as vertical, roughly tabular schlieren trending north-east, that is parallel to the trend of the adjacent ultrabasics.
2 The texture varies from equigranular in the coarse or

pegmatitic varieties to sub-porphyritic or ophitic in the microgabbros.

3 The effects of dynamic metamorphism are indicated by varied development of schistosity, brecciation and saussuritisation of the plagioclase feldspar. The gabbro is highly brecciated and sheared in the Malango area west of the Suta Ultrabasics outcrop, but is remarkably fresh where it intrudes the basalts of the Valasi Block.

The Mbirao Metabasics

A belt of metamorphosed basic rocks forms the major east–west 'axial zone' of eastern Guadalcanal, which extends from the upper Sutakama Valley in south-central Guadalcanal to Marapa Island, 85 km to the east. The belt narrows from 15 km in width in the west to only 1.5 km in the Marau area. Two smaller areas of highly altered, sheared and brecciated basic igneous rocks have also been classified here as Mbirao Metabasics: an area of about 17 km² in the Mbeambea Valley and a further 7 km² in the Mbetilonga Basin.

The Rere River and its headwater the Tiraone provide the most revealing section through the Mbirao Metabasics. The western boundary of the main outcrop of the Mbirao Metabasics is formed in part by the line of the north-easterly-trending Suta Ultrabasics; the southern end of this body curves round to the south-east along the Kuma Valley, following the line of the Suta Fault, which brings the Lower Miocene Suta Volcanics on the south-west side into juxtaposition with the metabasics to the north-east.

To the north the metabasics are overlain unconformably by Lower Miocene–Pliocene calcarenites. The surface of unconformity may be examined in the Kolokumaha and Vaghanambo river sections: elsewhere it is not exposed or has been faulted out.

Except along the line of the Mandonu Fault, at the eastern end of the outcrop, there is no sharp break in rock type between the Weather Coast Block and the metabasics to the north. On traversing the watershed from south to north one gains the impression of a gradual transition from relatively unsheared basalts and subsidiary dolerites to altered, brecciated and schistose metabasics; northwards the schistosity becomes more pronounced, until the metamorphic effect penetrates the entire outcrop in the axial zone.

The rivers tend to follow fault-lines, and thus the impression gained from river traverses may be that dynamic metamorphism is more widespread than is actually the case, although long reaches may be observed in which the strike of the s-surfaces is oblique to the general trend of the valleys, for example in the Kolopao tributary of the Lughumboko River. The disposition and significance of the metamorphic s-surfaces are discussed in some detail in Chapter 3.

The following major rock-types may be distinguished, although every gradation can be observed from one to another, even over a distance of 10 m:

Unsheared basaltic lavas.

Unsheared gabbro or microgabbro.

Brecciated lava.

Flaser rocks, in which sufficient of the original mineralogy and texture remains to assess whether the parent rock was a coarse intrusive or a lava.

Undifferentiated mylonites, appearing in the field as tough flaggy greenish-grey rocks with a crude foliation.

Well-foliated rocks, such as phyllonite and chlorite-actinolite-schists, which weather to a rusty blue-grey shale.

Shear breccias: a group of polygenetic rocks which may represent sheared autobrecciated basalt, brecciated schist, fault breccia, etc. Cataclasis has reduced such rocks to a common end-product.

Epidosites: rocks composed entirely of minerals of the epidote group, together with secondary quartz.

Subsidiary lenses of limestone, recrystallised and sheared, and ferruginous chert.

Minor occurrences of carbonaceous schist (Plate 7) with magnetic segregations and metallic sulphide lodes. These are known largely from floaters.

There is some evidence that the metamorphic grade increases eastwards, that is towards the narrow end of the metabasic belt in the Mbirao Valley Zone; particularly in the Kolovaghamela and Mandonu rivers, there are zones of chlorite-actinolite-schist, in part associated with the margins of the Marau Ultrabasic body; it was to these rocks that Thompson (1960) first applied the name Guadalcanal Schists.

Plate 7 Photomicrograph of graphite-schist from the Mbirao Metabasics. Mborambora River, south-central Guadalcanal. Plane polarised light.
Folia of chlorite and sericitic mica are folded with flaky metallic graphite which has partly migrated into an axial slip. There are some very finely divided granules of clinopyroxene

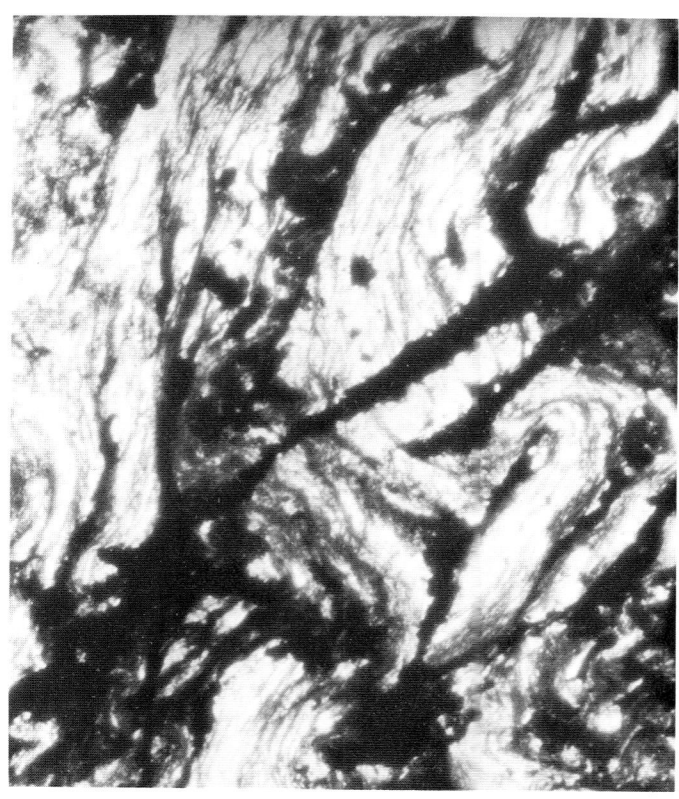

Table 2 Chemical analyses of basalts from the Mbirao Group

	1	2	3	4	5
SiO_2	49.82	47.2	47.63	48.3	48.73
Al_2O_3	13.12	12.8	14.15	14.6	14.66
Fe_2O_3	2.56	3.9	1.90	1.4	1.58
FeO	11.39	9.7	8.61	8.2	7.99
MgO	5.88	6.0	8.12	8.7	8.10
CaO	10.17	9.4	12.62	12.2	12.36
Na_2O	2.22	5.1	2.36	2.6	2.12
K_2O	0.21	0.17	0.07	0.05	0.05
H_2O+	2.57	1.8	1.77	1.5	1.76
H_2O-	0.06	0.22	0.21	0.1	0.14
TiO_2	1.56	2.1	1.23	1.45	1.44
P_2O_5	0.10	0.28	0.12	0.19	0.14
MnO	0.23	0.38	0.20	0.31	0.19
CO_2	n.d.	n.d.	1.24	n.d.	0.26
	99.89	99.05	100.23	99.60	99.52
		mg/kg		mg/kg	
Cu		60		100	
Ni		110		290	
Co		70		100	
NORMS					
Q	3.56	—	—	—	—
Or	1.23	1.00	0.39	0.28	0.28
Ab	18.78	30.00	19.98	21.98	17.94
An	25.23	11.54	27.82	28.03	30.35
Ne	—	7.13	—	—	—
Di Wo	10.27	13.88	14.21	13.07	12.27
En	4.75	7.48	8.33	7.93	7.42
Fs	5.41	5.92	5.19	4.42	4.18
Hy En	9.98	—	3.82	3.56	11.63
Of	11.24	—	2.37	1.99	6.56
Ol Fo	—	5.23	5.66	7.12	0.77
Fa	—	4.57	3.87	4.38	0.49
Mag	3.71	5.65	2.76	2.02	2.29
Il	2.96	3.99	2.34	2.76	2.73
Ap	0.24	0.67	0.27	0.44	0.57
C.I.	43.59	33.90	54.11	55.89	55.27
A'	—	—	+8.96	+12.03	−14.62
$\dfrac{Al_2O_3-Na_2O}{TiO_2}$	6.99	3.67	9.59	4.62	8.71

1   Quartz-tholeiite (65534), Sambaharihi River, 9° 51′ S, 160° 29′ E. Analysis by courtesy of R. L. Stanton
2   Alkali-basalt (65507), Nggeunaha River, 9° 40′ S, 160° 15′ E. Analysis by Sheen Laboratories Pty Ltd, Western Australia
3   Olivine-tholeiite (BSI 6640), Kandaho River, 9° 53′ S, 160° 41′ E. Analysis by courtesy of P. Jakes
4   Olivine-tholeiite (65508), Tetevasa River, 9° 39′ S, 160° 28′ E
5   Alkali-olivine-tholeiite (BSI 7319B), Tetevasa River, 9° 39′ S, 160° 28′ E. Analysis by courtesy of P. Jakes

Sheared basalts are commonly pale bluish-green in colour, the plagioclase having been saussuritised and the pyroxenes converted to chlorite and actinolite. Weathered surfaces are often reddish-grey or light brown, due to the release of ferric iron from finely disseminated sulphide. Pillows may have survived the early stages of shearing, the cores of the pillows remaining as augen-like structures up to 1 m in diameter, surrounded by a sheared matrix.

Feldspar is rarely determinable in thin sections of sheared basalt, but poeciloblastic albite is clearly developed in the chlorite-actinolite-schists. Vein quartz often shows strain shadows, and pale green actinolite commonly forms veinlets of cross-fibre type.

The Mbirao Metabasics are the metamorphosed equivalents of other representatives of the Mbirao Group. The metamorphism was essentially low-grade regional or dynamic and, from the occurrence of albite, clinozoisite and actinolite, may be classed in the greenschist facies of Turner (1968, p.268). A transition southwards to the pumpellyite and zeolite facies is suggested by the occurrence of those minerals in the basalts of the Weather Coast Block.

Petrochemistry of the Mbirao Group

Chemical analyses of ten rocks from the Mbirao Group are shown in Tables 2, 3, together with CIPW norms. In view of the widespread alteration of the Mbirao rocks, particularly of the volcanics, it would seem hazardous to generalise on the basis of a very small number of analyses. However, the volcanics and minor intrusives appear to share certain features in common; analyses from the Mbirao Group plot in distinct fields on each of the variation diagrams; in general these fields do not overlap with those of the Lower Miocene and younger volcanics (Figures 28 to 30 and 32).

In terms of the generalised normative tetrahedron of Yoder and Tilley (1962) the lavas may be classified as follows:
Oversaturated tholeiite, with normative quartz and
  hypersthene (65534).
Olivine-tholeiites, with normative hypersthene and olivine
  (65506, BSI. 6640, 65538, BSI. 7319B).
Alkali basalts (or olivine-basanite in the classification of
  Green, 1969), with normative olivine and nepheline
  (65507, BSI. X1413).
Olivine-tholeiite (65514—a basalt dyke from the Valasi
  Block).
Alkali olivine-basalt (65505—the matrix of the
  glomeroporphyrite dolerite).

Poldervaart (1964) used the index A' = Ab − 1.96 Enhy − 1.49 Ofhy to subdivide the olivine-tholeiites: alkali basalts give a negative value, tholeiites a positive. Accordingly three of the olivine-tholeiites would be classified as alkali basalts; however, although 65538 is an alkali basalt by this criterion, sample BSI 7319B, from the same flow, is a tholeiite.

There is a small range of silica percentage (46–50), but a relatively greater variation in the alkalis. These features are illustrated by the silica-variation diagram where the analyses of the Mbirao rocks plot on either side of the boundary between the fields of the tholeiitic and alkali series (Macdonald, 1960).

Table 3 Chemical analyses of schist, dolerites, microgabbro and leucogabbro from the Mbirao Group

	1	2	3	4	5
SiO$_2$	47.9	49.5	48.8	46.98	47.7
Al$_2$O$_3$	13.3	15.2	13.5	16.24	22.0
Fe$_2$O$_3$	2.6	0.4	2.2	trace	0.7
FeO	10.3	7.7	9.8	4.56	2.4
MgO	7.6	9.7	7.4	12.24	8.2
CaO	10.2	11.7	11.1	14.72	14.4
Na$_2$O	3.4	2.6	3.5	1.43	2.9
K$_2$O	0.1	0.06	0.19	0.45	0.03
H$_2$O +	1.8	1.4	1.7	2.81	1.8
H$_2$O −	0.2	0.2	0.1	0.12	0.1
TiO$_2$	1.3	0.54	1.3	0.24	0.16
P$_2$O$_5$	0.16	0.07	0.18	—	0.07
MnO	0.28	0.25	0.34	—	0.12
	99.14	99.32	100.11	99.79	100.58
	mg/kg	mg/kg	mg/kg		mg/kg
Cu	220	320	210		60
Ni	140	280	240		300
Co	60	90	90		70
NORMS					
Q	—	—	—	—	—
Or	0.61	0.33	1.11	2.67	0.17
Ab	28.79	22.03	27.85	8.55	15.16
An	20.72	29.63	20.55	36.56	46.92
Ne	—	—	0.97	1.93	5.11
Di Wo	12.05	11.67	13.92	15.23	10.04
En	6.47	7.09	7.49	10.97	7.60
Fs	5.17	3.93	5.96	2.88	1.41
Hy En	6.24	4.73	—	—	—
Of	4.98	2.63	—	—	—
Ol Fo	4.14	8.64	7.66	13.67	8.98
Fa	3.64	5.28	6.72	3.95	1.83
Mag	3.77	0.58	3.20	—	1.01
Il	2.47	1.03	2.48	0.46	0.30
Ap	0.36	0.17	0.44	—	0.17
C.I.	43.42	58.18	45.11	73.89	73.25
A′	+ 9.14	+11.06	—	—	—
$\dfrac{\text{Al}_2\text{O}_3 - \text{Na}_2\text{O}}{\text{TiO}_2}$	7.62	—	—	—	—

1 Chlorite-actinolite-schist (65506), Alivaghato River, 9° 45′ S, 160° 21′ E. Analysis by Sheen Laboratories Pty Ltd, Western Australia
2 Dolerite (65514), middle Vaghanambo River, 9° 38′ S, 160° 29′ E. Analysis by Sheen Laboratories Pty Ltd, Western Australia
3 Matrix of glomeroporphyritic dolerite (65505), upper Tanggiata River, 9° 49′ S, 160° 28′ E. Analysis by Sheen Laboratories Pty Ltd, Western Australia
4 Microgabbro (BSI X1413), Pina River, 9° 39′ S, 159° 46′ E. Analysis from Thompson (1960)
5 Leucogabbro (65510), Mburumburu, 9° 43′ S, 160° 40′ E. Analysis by Sheen Laboratories Pty Ltd, Western Australia

Sample 65506 (Table 3) is a chlorite-actinolite-schist from the Mbirao Metabasics, chemically not significantly different from its basaltic 'parents'.

The Mbirao Group analyses are plotted on the variation diagram used by Murata (1960) to distinguish between tholeiitic and alkalic basaltic series in Figure 32; the bias would appear to be in favour of the tholeiitic trend, but some of the analyses conform more closely to the trend of the alkalic series.

Analyses of spilitic rocks show considerable variations even within individual pillows (Vallance, 1960). Furthermore sampling of representative rocks for analysis presents considerable difficulties (Amstutz, 1968): the Mbirao samples were selected as being representative of the fresher parts of the outcrop, highly epidotic or oligomineralic facies being rejected.

Many more analyses of the Mbirao Volcanics are required before the petrochemical range can be adequately compared with that of other 'greenstone' complexes. Na$_2$O and SiO$_2$ appear to be somewhat lower than the average composition of Ural spilite (Na$_2$O 3.64 per cent, SiO$_2$ 52.45 per cent) given by Sharfman (1968); TiO$_2$ is much higher in the Mbirao rocks than Sharfman's average of 0.66 per cent; however Turner and Verhoogen (1951) regard a relatively high percentage of TiO$_2$ as a feature of the spilite–keratophyre association. The range of composition of the Mbirao rocks compares fairly well in many respects with analyses of so-called spilitic suites. The wide 'spread' on the silica-variation diagrams (Figure 28) precludes the possibility of obtaining a well defined alkali–lime index (Peacock, 1931); the intersection of the predominant trends could range from about 49 to 54 per cent SiO$_2$, indicative of an alkalic or alkali-calcic suite.

On an 'F'MA diagram (Figure 29) the Mbirao Volcanics plot along the trends of both the Hawaiian alkalic and tholeiitic series. The variation in alkali proportions could be a function of differential metasomatic effects subsequent to submarine extrusion at considerable depths; under these circumstances primary differentiation trends would tend to be somewhat obscured.

The percentage of titania is relatively high in the Mbirao Volcanics, which is interesting in view of the distinction of Chayes (1964, 1965) between oceanic and circumoceanic basalts on the basis of titania and alumina contents.

Grasso (1968) found that the index (Al$_2$O$_3$ − Na$_2$O)/TiO$_2$ varies from between 3.6 and 9.6 for the volcanics, suggesting an oceanic association (Atlantic in the older terminology) rather than a suite typical of orogenic environments.

THE GUADALCANAL GABBRO   One analysis of the Guadalcanal Gabbro (sample 65510) from the Mburumburu River in eastern Guadalcanal shows several anomalous features—in particular high percentages of Al$_2$O$_3$ and CaO. It is essentially a 'high-alumina' alkali-gabbro. In thin section the first impression is that the rock is anorthositic: subordinate pale green uralitised pyroxene is inconspicuous and there is as much as 75 to 80 per cent labradorite–bytownite. Moreover the MgO content is higher than that of the average anorthosite, mafic minerals comprising 31 to 34 per cent of the rock; according to the classification of Buddington (1939) the rock would be termed an anorthositic gabbro.

Another interesting feature is the very low TiO_2 content (0.16 per cent), in contrast to the other basic pre-Miocene rocks. In this respect it may be significant that a basaltic dyke intrusive into the gabbro of the Valasi Block (65514) is also very low in TiO_2; however in other respects this rock resembles an alkali-olivine-tholeiite, comparable chemically with the Mbirao Volcanics. An analysis of a late-state gabbro-pegmatite associated with the Guadalcanal Ultrabasics (Table 4) gave a value of 3.4 per cent TiO_2; possibly the upper part of the gabbro magma was depleted of pyroxene, iron oxides and titania by gravitational settling, and more mafic material at depth was tapped at the time of ultrabasic emplacement.

Table 4 Chemical analyses of rocks from the Guadalcanal Ultrabasics and their associates

	1	2	3
SiO_2	53.4	43.5	40.06
Al_2O_3	2.6	9.0	8.29
Fe_2O_3	1.4	6.2	3.73
FeO	6.3	14.5	6.92
MgO	25.7	8.6	29.24
CaO	10.0	10.2	5.35
Na_2O	0.1	1.9	0.64
K_2O	0.01	0.21	0.22
H_2O+	0.1	2.2	5.38
H_2O-	0.1	0.2	0.65
TiO_2	0.1	3.4	0.22
P_2O_5	0.07	0.15	—
MnO	0.31	0.32	—
	100.19	100.38	100.70

	mg/kg	mg/kg	
Cu	80	30	
Ni	1600	210	
Co	120	100	

NORMS

	1	2	3
Q	—	—	—
Or	0.06	1.22	1.28
Ab	0.84	16.10	5.40
An	6.62	15.41	19.11
Ne	—	—	—
Di Wo	17.76	14.27	3.10
En	13.59	7.78	2.44
Fs	2.30	5.99	0.32
Hy En	43.02	7.91	2.93
Of	7.27	6.08	0.37
Ol Fo	5.16	4.01	47.27
Fa	0.96	3.40	6.60
Mag	2.04	8.98	5.42
Il	0.20	6.46	0.42
Ap	0.17	0.37	—
C.I.	71.24	41.74	73.69

1 Websterite (65509), Charihaviha River, 9° 44′ S, 160° 07′ E. Analysis by Sheen Laboratories Pty Ltd, Western Australia
2 Gabbro-pegmatite (65-504), Charivegheu River, 9° 40′ S, 159° 52′ E. Analysis by Sheen Laboratories Pty Ltd, Western Australia
3 Troctolite (BSI X1409), Pina River, 9° 41′ S, 159° 46′ E. Analysis from Thompson (1960)

CONCLUSIONS The Mbirao Volcanics and associated hypabyssal rocks appear to belong to a suite of very altered alkali olivine-basalts and tholeiites occupying a broad field which overlaps to some extent the trends of the Hawaiian Alkalic and Tholeiitic Series.

Petrographically there are some features characteristic of the Spilite–Keratophyre Association but chemically there are affinities with the Hawaiian Oceanic Olivine-Basalt Association (Turner and Verhoogen, 1951). The variation in silica percentage is small, but there are considerable differences in the proportions of alkalis within the suite.

According to Joplin (1968, p.150) 'it seems likely that the "spilite suite" may be divided into two classes, rocks containing zeolites and other deuteric minerals and not associated with geosynclinal sediments, and rocks containing chlorite, epidote, albite and sometimes pumpellyite and hydrogrossular, but no zeolite, associated with geosynclinal sediments'.

The 'spilitic' features of the Mbirao Volcanics put the group with the former class: it is considered that they comprise a sequence of 'oceanic' alkali-basalts which have suffered deuteric alteration, and that the variations in the alkalis are partly the result of post-extrusive metasomatic effects.

One analysis of the Guadalcanal Gabbro shows some anorthositic tendencies; more analyses are needed to determine whether this is a localised feature, characteristic of the gabbros of the Valasi Block and the Kaoka Bay hinterland.

The $K_2O:Na_2O$ ratio appears to be uniformly low; the average value for the eight analyses from within the Mbirao Group is 0.03.

Normative proportions suggest that the plagioclase is in the labradorite range, varying to basic oligoclase in the case of 65507. The leucogabbro 65510 gives a bytownitic ratio, 76 per cent An, which accords with micropetrographic determinations.

THE GUADALCANAL ULTRABASICS

The ultrabasic rocks of Guadalcanal outcrop in three roughly linear belts:
1 The *Marau Ultrabasics* in the east form a discontinuous line of outcrops for 35 km from the Mbahara River east-south-east to the island of Maruiapa in Marau Sound.
2 The *Suta Ultrabasics* trend north-east for 16 km, flanking the east side of the Sutakama Valley in central Guadalcanal.
3 The *Ghausava–Itina Ultrabasics* in the south-west occupy a broad, discontinuous zone trending for 25 km east-south-east of Beaufort Bay. Unpublished information on the Marau and Ghausava Ultrabasics is available in Thompson (1960).

Areas of ultrabasic rock are distinctive photogeologically, since they give rise to light-toned areas of heath or sparse forest with rounded contours, the 'cushion' topography of Stanton (1961). However, the narrow Suta belt does not exhibit these features as distinctively as the Marau or Ghausava bodies (Plate 2).

Coleman (1965a) considered that the ultrabasics of Guadalcanal were emplaced in the Upper Oligocene, because field relationships suggest that they are younger than the basement lavas, and their erosional detritus, first found in the Aquitanian, is common only in Burdigalian and later sediments.

Thompson (1960) considered that the emplacement of the Ghausava Ultrabasics commenced in the Oligocene and continued until after the Miocene; however he suggested a Middle or Upper Miocene age for the emplacement of the Marau Ultrabasics since the brecciation and recrystallisation of the Lower Miocene Lake Lee Calcarenite overlying the Kaoka Lavas (= Mbirao Volcanics) was believed to be associated with the ultrabasic emplacement. The writer's mapping has since shown that the brecciated limestones which outcrop adjacent to the Marau Ultrabasics must be correlated with the pre-Miocene Tetekanji Limestones: Miocene limestone does not occur east of Mt Vatupochau.

Since the Marau and Suta Ultrabasics have been affected by the same stresses which impressed a sequence of schistosites on the surrounding Mbirao Metabasics, it may be inferred that they were first emplaced in late 'Mbirao' time, that is the Eocene or possibly Oligocene. Figure 3 indicates tentatively initial emplacement in the Eocene and remobilisation in the Oligocene. Structural evidence, detailed in Chapter 3, indicates that pre-Miocene rocks have locally been thrust over Lower Miocene sediments; it is therefore probable, in view of the mobility of serpentinite, that the ultrabasics were thrust higher into the crust in the late Miocene or Pliocene. In the case of the Suta Ultrabasics, Pliocene conglomerates rest nonconformably on serpentinite, but the boundary with Miocene sediments has been tectonised. However the structure of the Ghausava–Itina Ultrabasics, mapped by Thompson (1960, 1968), cannot yet be interpreted in the light of detailed structural information from the Suta and Marau occurrences.

The Marau Ultrabasics

Good exposures are found in the lower tributaries of the Kolovaghamela River, notably the Vaumandali, Vaitambu and I Mbosa.

The ultrabasic belt shows some deviation from a strictly linear form, in part due to dextral transcurrent faulting along north-westerly lines. The serpentinite has been thrust into basement metavolcanics along most of the outcrop, although Guadalcanal Gabbro is in contact with ultrabasics for 15 km along the latter's northern boundary.

Serpentinised harzburgite makes up 90 per cent of the entire outcrop; freshly hammered surfaces are greenish-black and vitreous, with 'eyes' of bastite (after enstatite) prominent in a mesh-like matrix of serpentine minerals. Weathered surfaces often show a multicoloured 'checkered' effect in pinkish brown, green and greyish black. There is much evidence for shearing which postdated the serpentinisation: slickensided surfaces are common, and early shears may be dragfolded or transected by later shears. Locally veinlets of copper oxide, malachite and hematite have been found.

In thin section (Plate 8) islands of colourless altered enstatite up to 4 mm in diameter give a pseudoporphyritic effect to an otherwise hypautomorphic-granular texture; relict cores of olivine are surrounded by a meshwork of fibrolamellar channels. Some of the original enstatite–olivine rocks had less than 5 per cent iron ore in the form of mag-

netite, chromite or chromian hercynite; accordingly they are classified as saxonites (Johannsen, 1939, Vol. 4, p. 435). Some harzburgites show minor accessory diopsidic augite, but lherzolites, with higher proportions of clinopyroxene, have not been recognised.

Minor dykes of pyroxenite do occur, however.

ULTRABASIC ASSOCIATES   Rock types differing from the normal serpentinised harzburgite are exposed over about 1 km of the middle Kolovaghamela gorge, and in the upper part of the I Mbosa tributary. In the gorge of the main river the serpentinite has been intimately penetrated by veinlets of white plagioclase and pyroxene ranging in width from 1 mm to 1 cm. The trend of the veins is variable; locally they trace out angular ptygmatic folds whose axial planes strike north-east and dip at 70° to the south-east. Every gradation is observed between serpentinised harzburgite intruded by minor veins of pyroxenite and anorthosite, and larger bodies of gabbro; sheared intermediate types resemble dark green amphibolites with porphyroid 'eyes' of feldspathic material, relics of original veining.

In thin sections, the ptygmatic veins and relict 'eyes' are monomineralic, consisting entirely of either pale brown pleochroic orthopyroxene or bytownitic plagioclase.

Plate 8   Photomicrograph of harzburgite from the Marau Ultrabasics, I Mbosa River, east Guadalcanal. Crossed nicols.
Islands of enstatite are surrounded by channels of pale green fibro-lamellar antigorite. Accessory magnetite. Texture hypautomorphic-granular

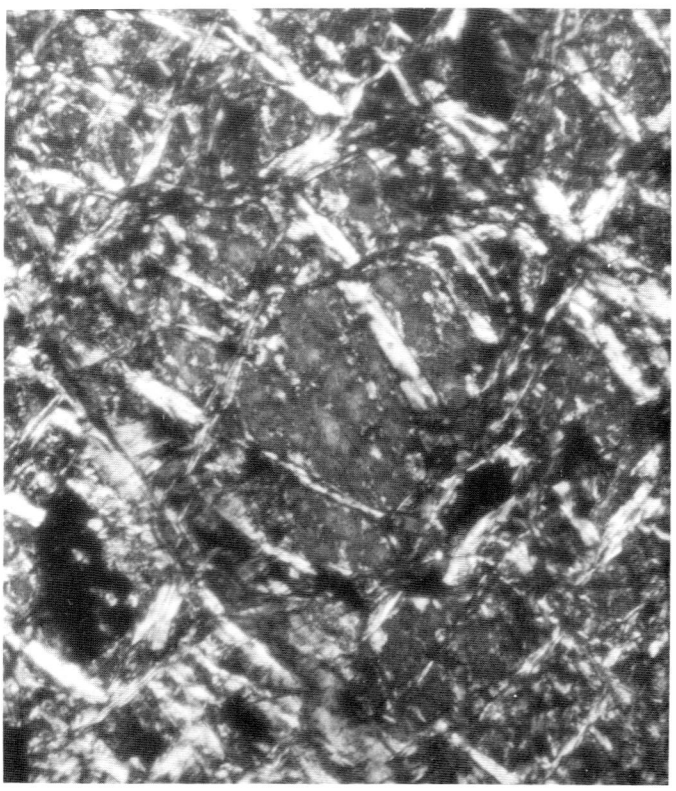

In the I Mbosa River, a dyke-like body intruding serpentinised harzburgite is an amphibolite or albite-hornblende-gneiss, composed of almost equal proportions of hornblende and plagioclase. The hornblende occurs as ragged porphyroblasts up to 2 mm in length, enclosing minute anhedral xenocrysts of epidote, sphene, iron ore, etc. Occasionally relict pyroxene structure is preserved. This hornblende is strongly pleochroic from deep blue or sea-green through yellow-green to colourless or neutral; according to Shidô and Miyashiro (1959) such bluish-green hornblende is typical of a grade of metamorphism intermediate between an actinolitic facies and a higher-grade zone characterised by greenish-brown hornblende.

A similar type of albite-actinolite-schist forms a dyke one metre wide intrusive into the Guadalcanal Gabbro 100 m north of the boundary with the serpentinite in the Vaumandali River.

The metamorphic grade of these rocks must be classed with the amphibolite facies of Turner (1968), that is higher than the surrounding greenschist of the Mbirao Group. Slices of metagabbro intricately associated with serpentinite have been described from other ultrabasic areas in the Solomons, for example, San Jorge (Marshall and others, 1957). Presumably these rocks have been carried up with the serpentinite from a deeper metamorphic zone.

Floaters of olivine-gabbro have also been found in the lower Kolovaghamela River, but are not mapped in situ; as olivine has not been recorded in the Guadalcanal Gabbro, it is presumed that these boulders have been derived from minor bodies associated with the Marau Ultrabasics.

The Suta Ultrabasics

The Suta Ultrabasics (new name), named for that district of Guadalcanal which is drained by the Mbalisuna headwaters, outcrop over an area of about 14 km². Good exposures are available in the tributaries which drain the right side of the Sutakama Valley, and in the upper Kuma Valley. The main portion of the outcrop, at the most 2 km in width, extends north-east for 13 km along the eastern flank of the Sutakama Valley: at the southern end, the outcrop becomes discontinuous, swinging round to the south to join an arcuate belt of smaller separated bodies which change trend from north–south in the Sutakiki Valley to north-west–south-east in the Kuma headwaters. With the possible exception of eastern Choiseul, the larger ultrabasic bodies of the Solomons are arranged in linear fashion parallel to the north-westerly trend of the major islands, so that the Suta occurrence constitutes an important anomaly in this pattern.

The main part of the Suta Ultrabasics appears to have been thrust into place along the faulted boundary between Mbirao Metabasics on the east side and the Guadalcanal Gabbro to the west; either faulting pre-dated the ultrabasic emplacement, or else the gabbro and ultrabasics were thrust over the metabasics and then later the ultrabasics were shifted higher in the crust.

The ultrabasic contacts are faulted except at one locality: in the right bank of the Sutakama River just below its confluence with the Mbongialo tributary, a serpentinised harzburgite dyke about 30 m in width and trending north–south,

contains irregular lenses of gabbroic material. On the eastern contact is a zone of garnet-mica-schist 2 m wide which grades eastwards into sheared augite-gabbro typical of the pre-Miocene basement. Conspicuous red almandine garnets attain a diameter of 4 cm, and the schistosity is parallel to the margins of the serpentinite dyke. Since the amount of heat produced by serpentinisation would have been inadequate to form almandine, it must be assumed that the schists were contact-metamorphosed by hot harzburgite, the almandine developing at the expense of chlorite (compare Tilley, 1926). As in the Marau Ultrabasics, the predominant rock type is a serpentinised harzburgite. However, there are occasional dykes and veins of gabbro and pyroxenite.

In the Sutakama River 2 km below Nuhu village the serpentinite has been intricately veined by feldspathic and pyroxenitic material. Veins up to 11 cm in width trend roughly east–west and dip steeply north: they have been folded in a manner suggesting the influence of east–west sinistral transcurrent movement while the ultrabasics were still in a plastic state.

In the Kuma headwaters and the Charihaviha tributary of the Sutakama, the serpentinite has been veined by anastomosing dyke-like bodies of coarse leucocratic enstatite-gabbro (Plate 9). Pods of coarse serpentinite may be enveloped in the gabbro, but the converse relationship is also found, where islets of gabbro appear to be surrounded by ultrabasic material. The contacts between serpentinite and gabbro are sharp with no evidence for chilling. Generally the gabbro dykes trend parallel to the strike of the enclosing ultrabasics. A sample of leucogabbro, from the Charihaviha River, shows grey prismatic enstatite crystals ($2V \approx 80°$, optically +ve) with very narrow reaction rims of finely granular clinopyroxene. The plagioclase is bytownite (An_{80}). Both pyroxene and feldspar show strain shadows, probably due to protoclastic deformation.

Minor pyroxenite dykes may be websteritic, with roughly equal proportions of diallage and enstatite or composed entirely of bronzitic orthopyroxene.

Varieties of dunite-serpentinite and sheared troctolite have been found associated with the smaller ultrabasic bodies outcropping to the west of the Sutakama River.

The Ghausava–Itina Ultrabasics

The Ghausava body was described by Thompson (1960), who also discovered the Itina ultrabasics when mapping was extended eastwards (Thompson, 1968).

Although essentially parallel to the Marau Ultrabasics, the crescentic Ghausava body was emplaced into the pre-Miocene basement up a southerly-dipping thrust plane. Furthermore, the Guadalcanal Gabbro outcrops on the south side, whereas the basement lavas, here equated with the Mbirao Volcanics, are exposed in the Ghausava River on the north side of the ultrabasics. These relationships are the reverse of those obtaining for the Marau Ultrabasics, which are inclined to the north, but compare better with the structure of Santa Isabel (Thompson, 1960).

The bulk of the Ghausava–Itina mass is made up of serpentinised harzburgite with subsidiary intrusions of gabbro-pegmatite. There is no evidence for contact metamorphism.

The possibility of other ultrabasic areas

Boulders of amphibolite, gabbro with melanocratic schlieren, and serpentinite have been found in the Mberande and Mbokokimbo rivers. None of these indicators have been traced to an 'in situ' source: possibly they have been derived secondarily from clasts in Pliocene conglomerates but, on account of exposure failure, small ultrabasic areas could still remain unmapped in the Mbokokimbo headwaters.

Petrochemistry of the ultrabasics

On the basis of visual estimations, many of the partially serpentinised ultrabasic rocks from the Marau and Suta outcrops carry over 70 per cent orthopyroxene. Sample 65509, from the southern end of the Suta Ultrabasics, superficially typical of much of the outcrop, is in fact a websterite in the classification of Malakhov (1964) with over 50 per cent normative hypersthene and 33 per cent normative diopside. Silica is much higher than in the harzburgites as originally defined (Johannsen, 1939, Vol. 4, p. 439).

An analysis of gabbro-pegmatite (sample 65504, collected by R. B. Thompson), is also shown in Table 4, together with an analysis of troctolite from the Pina River, west Guadalcanal (Thompson, 1960). Both rock types are associated with the Ghausava–Itina ultrabasic mass of western Guadalcanal; the analyses illustrate an interesting contrast, in that the gabbro-pegmatite, which intimately veins harzburgite, is considerably enriched in iron, whereas the weight per cent MgO in the troctolite (a discrete body) is nearly four times that for the pegmatite. As the ultrabasics are demonstrably the youngest phase of the pre-Miocene complex, it may be inferred that iron has been retained in the late-stage differentiates.

The Suta Volcanics

The name Suta Volcanics has been applied to a sequence of basalts, basaltic andesites, associated pyroclastics and subsidiary limestones which outcrop over an area of about 150 km^2 in south-central Guadalcanal Hackman, 1968c). A small inlier in younger sediments is also found in the gorge of the middle Tina River in north-central Guadalcanal.

On the grounds of similarity of rock type and stratigraphic position, two other rock units are tentatively correlated with the Suta Volcanics (see Geological map of Guadalcanal).
1 The Marasa Volcanics of Thompson 1960), which occur on the south-west coast of Guadalcanal between Cape Beaufort and Mbolonda Bay.
2 The andesites of the pre-Miocene basement of north-west Guadalcanal (Pudsey-Dawson and Thompson, 1958) which outcrop in the headwaters of the rivers between the Umasani and Poha, also referred to as the Umasani Volcanics.

The Suta Volcanics form the highest and most rugged terrain in Guadalcanal, building a horseshoe of mountains around the Koloula Valley, and including the summits of Mts Makarakomburu and Popomanaseu (Tetena Haiacha). On aerial photographs the outcrop appears massive, with a lighter tone than that of the Mbirao Volcanics. The drainage is coarsely dendritic and deeply incised. The coastal fringes are exceptionally steep and subject to landslip.

Exposures are not easily accessible, and nowhere is there a continuous section through the volcanic pile. The fresh volcanics may be studied in the very steep boulder-choked

Plate 9 Serpentinised harzburgite with dykes of anorthosite and enstatite-leucogabbro. Suta Ultrabasics, upper Kuma River, south-central Guadalcanal

tributary ravines of the bigger rivers, notably in the head-waters of the Sutakama, the Aliemba tributary of the Kuma and the smaller watercourses draining to the Malag-heti coast.

Limestones interbedded with the volcanics have yielded abundant non-diagnostic planktonic foraminifera, also *Le-pidocyclina sp.* and *Miogypsina thecidaeformis*. Spheroidal algal concretions, embedded in a matrix of tuffaceous calcilutite, are a feature of some outcrops in the upper Kuma Valley (Plate 10).

The benthonic foraminifera indicate a Lower Miocene age; however, in view of the thickness of the lava pile under-lying the intercalated limestones, it is probable that the Suta Volcanics are partly Oligocene.

The main outcrop of the Suta Volcanics occupies a tri-angular area bounded by the sea to the south and faulted on the north-west and north-east sides. Relationships with ad-jacent rock units are difficult to establish, particularly as the boundary fault-lines, except along the middle part of the Kuma Valley, traverse steep densely forested ranges, poorly exposed and accessible only at the expense of great effort. The central part of the outcrop has been invaded by a Plio-Pleistocene intrusive, the Koloula Diorite.

From fragmentary evidence in the middle Tina and upper Tinahulu rivers, it appears likely that the Suta Volcanics rest unconformably on Guadalcanal Gabbro or Mbirao Metabasics.

In the Sutalanga River section of the upper Itina, flows of basaltic andesite of Suta Volcanics type interfinger with Lower Miocene greywackes (the Kavo Greywacke Beds). To the north and north-east intercalated calcarenites indi-cate a transition in facies to the Lower Miocene Mbetilonga Limestones. In fact narrow zones of rapid facies change appear to coincide broadly with the lines of both the Mal-ango and Suta Faults. Measured dips of stratification are thin on the ground; however they tend to support the view that the folding which has affected the greywackes to the north-east (Thompson, 1968) has also involved the Suta Volcanics (Figure 13, sub-areas 8A and 8B).

The thickness of the Suta Volcanics, estimated from verti-cal sections, is at least 2500 m.

The Suta Volcanics show a much greater variation of rock type than the Mbirao Volcanics. Their petrography may be described under the following categories:
Feldsparphyric basaltic andesites.
Tualoto Limestones.
Vesicular pyroxene-basalts.
Pyroclastic rocks.

Feldsparphyric basaltic andesites are the most conspicuous lava type within the Suta Volcanics, and constitute perhaps 60 per cent of the total volume of the pile. Typical is a tough, massive, sparsely jointed mesocratic rock, fresh and resistant to weathering except where involved in shear zones related to major faults or where affected hydrothermally in the aureole of the Koloula Diorite. In hand specimen euhedral phenocrysts of white feldspar up to 8 mm in length are pro-minent in a greenish-grey groundmass; sometimes green clinopyroxene phenocrysts are also discernible. Quartz is not conspicuous in hand specimens except in the Koloula Valley where much of it has been introduced hydrother-mally in connection with the emplacement of the diorite.

In thin section, the plagioclase phenocrysts are commonly turbid, or have an outer turbid zone, and vary in composi-tion from oligoclase to labradorite. The feldspars often show a distinctive post-crystalline fissuring. Pale-green diopsidic augite $(2V \approx 60°)$ is subordinate to the feldspar. Occasion-ally sanidine has been identified, while accessory minerals (mostly magnetite) are not abundant; these features suggest that some of the flows may be of trachyandesite (Williams and others, 1954, p. 97). Hornblende is rarely seen in the suite and is mostly pseudomorphed by magnetite.

In the aureole of the Koloula Diorite, patches of quartz may occur as rounded grains or anhedral aggregates in the basaltic andesite: these may be associated with peculiar

Plate 10 Algal balls in a matrix of volcanic rudite, Suta Volcanics, upper Kuma Valley, south-central Guadalcanal

'clots' or intergrowths of biotite, actinolitic hornblende and iron ore.

Small volumes of other lava types are locally associated with the basaltic andesites, notably olivine-andesite and nepheline-trachyte.

The Tualoto Limestones may be considered as a sedimentary member of the Suta Volcanics division, intercalated within the basaltic andesites towards the north-east side of the Suta 'Triangle'; they outcrop in the Sutakama headwaters and the ravines draining the west side of the upper Kuma Valley. The commonest rock type is tough and grey where fresh, massive, and weathers to a brownish-grey colour with the superficial aspect of a dolerite. In thin section it is found to be composed of angular lithic fragments, predominantly of basaltic material, set in a calcareous matrix. The proportion of clasts to matrix is typically high, but varies from that of a lithic volcanic arenite with about 5 per cent calcareous matrix to a biomicrite contaminated with fine brown pelitic material, and composed largely of planktonic foraminifera.

Vesicular pyroxene-basalts outcrop in the south-west corner of the Suta Volcanics 'Triangle', overlying the basaltic andesites. They are well exposed in the Kologhailava and Ghalighecha rivers and in the coastal sections west of Nduindui village, where stratified flows of fairly massive basalt alternate with layers of pillow lava up to 50 m in thickness. The appearance varies with the relative size and abundance of white, green or pink zeolites and phenocrysts of zoned plagioclase or green pyroxene. In a representative thin section from the Kaimbaghasu River, 60 per cent of the phenocrysts are plagioclase with a compositional range of An_{60} to An_{75}. The feldspar is turbid, fissured and sericitised. Colourless or pale green subhedral diopsidic augite ($2V \approx 60°$) constitutes 30 to 40 per cent of the phenocrysts.

A considerable volume of *pyroclastic material* is spatially associated with the three categories described above. Fine-grained crystal tuffs do not bulk large; they are common in the lower part of the sequence, associated with the Tualoto Limestones. Lithic tuffs composed predominantly of basic lava fragments are commonly found in association with the feldsparphyric basaltic andesites. Volcanic breccias form perhaps 30 per cent of the middle and upper part of the Suta Volcanics. They are best described as volcanic breccias (Wentworth and Williams, 1932) in that most fragments are greater than 32 mm in diameter and are set in a coherent fine-grained tuffaceous matrix. The fragments are predominantly of basaltic andesite or vesicular basalt: accidental constituents are very rare.

Near the roof of the Koloula Diorite, these breccias may be highly indurated and silicified.

Occasional floaters found in the lower Koloula Valley are composed of light grey tuffaceous material with a eutaxitic structure, enclosing lenticles of dark vitric material with flame-like ends. Such structures are reminiscent of the 'tuff-lavas' of Bersenev and others (1966).

As most of the volcanic breccias are intercalated with massive flows of basalt and basaltic andesite, and have a restricted compositional range of components, it is probable that they are autoclastic flow breccias in the sense of Fisher (1960), resulting from the fragmentation of congealing lava during its advance.

The pillow formation and association with intercalated calcareous sediments suggest that the bulk of the Suta Volcanics was extruded in a marine environment, probably with fringing reef.

Petrochemistry

Seven analyses of rocks typical of the Suta Volcanics suite are presented in Table 5 together with norms.

The plots on the variation diagrams (Figures 28–30) indicate a wider range of composition than is present in the Mbirao Group, but a more distinct differentiation trend.

The rocks of the Suta Volcanics may be classified petrochemically as follows:

Rock type		Representative sample numbers
Quartz-normative tholeiites	Andesine-basalt	65511
	Labradorite-basalts	65535, 1455, 65515
	Bytownite-basalt	65536
Olivine tholeiite	Bytownite-basalt	65537
Alkali-olivine tholeiite	Andesine-dolerite	65508

Although the andesine-dolerite is an alkaline type, the other six are clearly tholeiitic in the sense of Yoder and Tilley (1962): samples 65511 and 1455 plot within the field of Kuno's High Alumina Basalt Series on a silica variation diagram (Figure 28) and within the field of the Japanese Hypersthenic Series on an 'F'MA diagram (Figure 30).

There is not a large range in silica percentage, but the suite differs from that of the Mbirao Group in the following respects:

1 There is a higher Al_2O_3 content than in the Mbirao rocks;

2 The average value for the K_2O/Na_2O ratio is 0.28; it is considerably higher than for the rocks of the Mbirao Group (0.03);

3 The titania content is lower than in the Mbirao Group and the average value for the index $\dfrac{Al_2O_3 - Na_2O}{TiO_2}$ is 21.00, which is more characteristic of an 'orogenic' or island-arc setting than of oceanic lavas according to Grasso (1968).

The zoning and alteration of the feldspars render difficult the determination of the plagioclase composition in thin section; normative proportions indicate a range from sodic bytownite in the olivine tholeiite to sodic andesine in the alkali-olivine-dolerite.

Intersections of the CaO and total alkali trends on the silica-variation diagrams give an alkali-lime index (Peacock, 1931) of approximately 62, that is indicative of a calc-alkalic series.

The Poha Diorite

A dioritic intrusive has been mapped in the Poha River and adjacent streams between the Lungga and Umasani rivers

Table 5 Chemical analyses of rocks from the Suta Volcanics

	1	2	3	4	5	6	7
SiO_2	49.7	58.3	53.04	48.42	50.6	49.3	51.6
Al_2O_3	14.2	14.6	18.62	16.06	14.5	16.5	17.8
Fe_2O_3	3.1	3.1	3.44	3.02	2.0	4.3	3.6
FeO	5.4	5.6	5.43	7.25	5.9	4.9	5.4
MgO	8.5	4.3	2.86	5.48	10.7	5.5	6.1
CaO	9.5	6.9	8.85	9.28	12.7	10.0	9.9
Na_2O	4.0	3.8	3.36	1.26	1.8	1.7	3.2
K_2O	0.67	0.77	1.40	0.15	0.64	0.72	0.81
H_2O+	1.8	1.5	1.23	4.12	0.6	5.6	0.5
H_2O-	1.2	0.2	0.03	0.05	0.1	1.4	0.1
TiO_2	0.56	0.74	0.86	0.58	0.50	0.60	0.70
P_2O_5	0.13	0.20	0.18	0.05	0.12	0.16	0.19
MnO	0.25	0.24	0.15	0.19	0.27	0.27	0.35
CO_2	n.d.	n.d.	1.29	3.72	n.d.	n.d.	n.d.
	99.01	100.25	100.74	99.63	100.43	100.95	100.25
	mg/kg	mg/kg			mg/kg	mg/kg	mg/kg
Cu	100	160			170	220	230
Ni	250	130			210	170	200
Co	70	40			90	70	70
NORMS							
Q	—	12.14	5.09	9.69	—	10.36	0.64
Or	3.95	4.56	8.29	0.89	3.79	4.23	4.79
Ab	31.74	32.15	28.43	10.65	15.21	14.37	27.06
An	18.83	20.50	31.58	37.73	29.60	35.30	31.82
Ne	1.14	—	—	—	—	—	—
Di Wo	11.36	5.18	4.65	3.32	13.65	5.54	6.72
En	7.86	3.00	2.45	1.84	9.41	4.79	4.39
Fs	2.56	1.94	2.06	1.36	3.13	—	1.86
Hy En	—	7.71	4.67	11.81	11.57	11.69	10.80
Of	—	5.00	3.93	8.85	3.85	—	4.58
Ol Fo	9.32	—	—	—	3.97	—	—
Fa	3.35	—	—	—	1.45	—	—
Mag	4.49	4.49	4.98	4.38	2.90	5.26	5.21
He	—	—	—	—	—	0.67	—
Il	1.06	1.41	1.64	1.11	0.96	1.14	1.34
Ap	0.30	0.47	0.44	0.13	0.27	0.37	0.44
C.I.	47.00	32.86	40.13	49.98	62.27	57.44	40.03
A′	—	—	—	—	−13.21	—	—
$\dfrac{Al_2O_3-Na_2O}{TiO_2}$	18.22	14.60	17.74	25.52	25.40	24.67	20.86

1 Alkali olivine-dolerite (65508), upper Kolonggulai (tributary of the Sutakama River), 9° 43′ S, 160° 07′ E. Analysis by Sheen Laboratories Pty Ltd, Western Australia
2 Quartz-tholeiite (65511), lower Koloula Valley, 9° 49′ S, 160° 04′ E. Analysis by Sheen Laboratories Pty Ltd, Western Australia
3 Quartz-tholeiite (65535), lower Koloula Valley above Kakake, 9° 49′ S, 160° 04′ E. Analysis by courtesy of P. Jakes
4 Quartz-tholeiite (65536), Kolomoli River, Malagheti, 9° 48′ S, 160° 06′ E. Analysis by courtesy of P. Jakes
5 Olivine-tholeiite (65537), Kolokoo River above Kololata, 9° 46′ S, 160° 01′ E. Analysis by Sheen Laboratories Pty, Western Australia
6 Quartz-tholeiite (65515), Kaimbaghasu River, 9° 47′ S, 159° 58′ E. Analysis by Sheen Laboratories Pty Ltd, Western Australia
7 Quartz-tholeiite (1455), Sisi River, 9° 48′ S, 160° 02′ E. Analysis by Sheen Laboratories Pty Ltd, Western Australia

(Thompson and Pudsey-Dawson, 1958). The diorite has been emplaced into hydrothermally altered augite-andesites here correlated tentatively with the Suta Volcanics. It is intruded by later andesitic dykes and overlain unconformably by the Lower Miocene Mbonehe Limestone. An age determination is needed to ascertain whether the Poha Diorite represents an acid phase of Oligocene–early Miocene activity broadly contemporaneous with that of the Suta Volcanics. Petrographically the rock is composed predominantly of hornblende and andesine, but quartz and biotite occur in a tonalitic variety.

Petrochemistry

One analysis of the Poha diorite (BSI 8199) is presented in Table 6; it may usefully be compared with three hornblende-andesites from the Plio-Pleistocene Gallego Lavas (Table 7). The three analyses are relatively high in Al_2O_3 and plot in the same general area on a series of variation diagrams; they might be considered to represent a common magma-type.

Although regarded by previous authors as part of the pre-Miocene basement, the Poha Diorite is clearly very different petrologically from the Guadalcanal Gabbro of the Mbirao Group; normative quartz is lower than for the granodiorite-tonalite of the Koloula stock (sample no. 65512).

The Poha Diorite appears to be a high-alumina, calc-alkaline dioritic intrusive: normative proportions indicate plagioclase of approximately An_{50}.

Table 6 Chemical analyses of Tertiary plutonic rocks

	1	2	3	4
SiO_2	60.51	68.6	46.5	46.4
Al_2O_3	18.98	14.1	13.5	14.6
Fe_2O_3	2.37	1.6	7.6	1.8
FeO	3.29	2.1	7.5	9.0
MgO	2.70	1.5	9.7	10.0
CaO	6.64	3.3	9.6	12.5
Na_2O	3.77	4.7	2.1	3.6
K_2O	0.69	0.96	0.35	0.03
H_2O+	0.48	2.2	0.4	0.9
H_2O-	0.07	0.1	0.2	0.1
TiO_2	0.46	0.29	0.96	1.4
P_2O_5	0.11	0.08	0.27	0.13
MnO	0.07	0.06	0.45	0.40
	100.14	99.59	99.13	100.86
		mg/kg	mg/kg	mg/kg
Cu		30	220	230
Ni		140	150	360
Co		20	90	130
NORMS				
Q	16.45	27.77	—	—
Or	4.06	5.68	2.06	0.17
Ab	31.88	39.76	17.78	14.01
An	32.19	14.55	26.37	23.59
Ne	—	—	—	8.92
Cor	0.24	—	—	—
Di Wo	—	0.52	8.15	15.70
En	—	0.31	5.80	9.61
Fs	—	0.18	1.62	5.20
Hy En	6.73	3.42	17.83	—
Of	3.46	1.98	4.97	—
Ol Fo	—	—	0.37	10.71
Fa	—	—	0.10	6.38
Mag	3.43	2.32	11.03	2.62
Il	0.88	0.55	1.82	2.66
Ap	0.27	0.20	0.64	0.30
C.I.	37.06	18.13	52.53	55.10

1 Poha Diorite (8199), 9° 26′ S, 159° 51′ E. Analysis by J. D. Bell
2 Koloula Diorite (65512), lower Koloula Valley, 9° 47′ S, 160° 03′ E. Analysis by Sheen Laboratories Pty Ltd, Western Australia
3 Koloula Gabbro (65513), Charichighi River, 9° 47′ S, 160° 03′ E. Analysis by Sheen Laboratories Pty Ltd, Western Australia
4 Mbalisuna Gabbro (BSI 6459), Mbeghovila River, 9° 33′ S, 160° 15′ E. Analysis by Sheen Laboratories Pty Ltd, Western Australia

The Kavo Greywacke Beds

As used here, the term Kavo Greywacke Beds is an extension of that used by Coleman (1965a) for 'feldspathic greywackes with foraminifera', of probable Lower Miocene age, exposed in the headwaters of the Mbetisahata River on the northern slopes of the Kavo Range. These rocks were first described by Hill (1960) as the Kavo Shales and Siltstones, but Coleman's term is preferred from petrographic considerations.

Greywacke beds outcrop in the Itina and Lamulaghi River basins; Thompson (1968) refers them to the Lungga Beds (Wright, 1968b). The writer has adopted Coleman's terminology, applying it to the Itina area on the grounds of contiguity, and also of stratigraphic and petrographic similarity to the original Kavo Greywacke Beds.

Although the rivers have deeply incised the terrain, the skyline of these sedimentary areas is less rugged than that of the Suta Volcanics to the east.

Table 7 Chemical analyses of rocks from the Gallego Lavas

	1	2	3
SiO_2	56.08	55.70	53.73
Al_2O_3	19.04	20.68	19.34
Fe_2O_3	3.70	4.19	5.24
FeO	3.10	2.17	3.06
MgO	3.83	2.46	3.56
CaO	8.04	7.63	7.80
Na_2O	4.18	4.19	2.82
K_2O	0.88	1.05	1.07
H_2O+	—	0.35	1.03
H_2O-	—	0.05	1.68
TiO_2	0.53	0.59	0.41
P_2O_5	0.16	0.22	0.17
MnO	0.14	0.13	0.15
Loss on ignition	0.82	—	—
	100.50	99.41	100.06
NORMS			
Q	6.10	8.35	11.41
Or	5.19	6.18	6.35
Ab	35.36	35.45	23.86
An	30.63	34.54	36.95
Di Wo	3.44	0.76	0.27
En	2.56	0.65	0.21
Fs	0.54	—	0.03
Hy En	6.98	5.47	8.66
Of	1.49	—	0.87
Mag	5.37	5.70	7.59
He	—	0.26	—
Il	1.00	1.12	0.77
Ap	0.37	0.19	0.40
C.I.	41.45	39.77	43.47
$\dfrac{Al_2O_3-Na_2O}{TiO_2}$	28.04	27.95	40.29

1 Hornblende-andesite (BSI J144), Isisu Point, 9° 15′ S, 159° 42′ E. Analysis by courtesy of P. Jakeš
2 Hornblende-andesite (65533), roadside 1.6 km east of Visale Mission, 9° 15′ S, 159° 42′ E. Analysis by J D. Bell
3 Hornblende-andesite (65503), tributary of Hanangsa River, Keli Ngali, 9° 55′ S, 160° 37′ E. Analysis by courtesy of R. L. Stanton

The best reference section is provided by the upper and middle sections of the Itina River. Vertical sections indicate a minimum thickness of 2500 m.

Occasional planktonic foraminifera, notably *Orbulina sp.*, are found in many thin sections. Coleman (1965a) considered these greywackes to be of probable Lower Miocene age.

In the upper Itina Valley the greywackes overlie and partly interdigitate with Suta Volcanics: they have been gently folded about axes trending north-west. To the north-east the Suta Fault defines the boundary with the Guadalcanal Gabbro. To the north-west, more detailed mapping is required to define the relationships with the Lungga Beds of Wright (1968b). On the west side a complex of faults trending north-north-east defines the margins of inliers of ultra-basics and pre-Miocene lavas of the Ghausava-Itina block.

Poorly sorted fine- and coarse-grained arenites alternate in thin beds; conglomeratic lenses are subordinate. The characteristic texture is that of a 'microbreccia', ranging in grain-size from the finest clay to 0.5 mm, exceptionally up to 1.5 mm. About 40 per cent of the larger fragments are of plagioclase, mostly andesine, showing the fissuring characteristic of the phenocrysts in the Suta Volcanics. Compositionally these greywackes are in fact comparable with the Suta basaltic andesites, and are presumed to be wholly derived from the erosion thereof: they are best described as volcanic wackes (Williams and others, 1954, p. 303). However occasionally the matrix of the wackes is highly calcareous.

Slump structures, graded bedding, small scale cross-bedding, load casts and discordant flammate contacts (Conybeare and Crook, 1968) are commonly observed in the greywackes.

The aspect of the Kavo Greywacke Beds is reminiscent of 'flysch' or the Greywacke Suite of Pettijohn (1954); they are here considered to have been deposited as marine turbidites derived from an unstable volcanic terrain.

The Mbetilonga Group

The name Mbetilonga Group is proposed for a sequence of biogenic limestones and calcarenites, predominantly of Lower Miocene age, but ranging into the Pliocene in eastern Guadalcanal. The group comprises the following formations:

Formation	Letter symbol	Age
Valasi Limestone	Tpv	Upper Miocene–Pliocene (Hackman, 1968c)
Mbonehe (Bonegi) Limestone	Tmh	Late Lower Miocene (Coleman, 1957)
Tina Calcarenite	Tmt	Late Lower Miocene
Lake Lee Calcarenite	Tml	Late Lower Miocene (Coleman, 1960b)
Mbetilonga (Betilonga) Limestone	Tmb	Early Lower Miocene (Coleman, 1960b)

The Mbetilonga Group outcrops discontinuously over a distance of 110 km between Mt Gallego in north-west Guadalcanal and Kaoka Bay in the east. Thompson (1968) and Wright (1968b) have mapped several small disconnected inliers in south-west Guadalcanal.

The outcrop of the Mbonehe Limestone extends to the west of the Lungga River: the Mbetilonga Limestone and overlying Tina Calcarenite outcrop discontinuously between the Lungga and Sutakama rivers in north-central Guadalcanal. The Lake Lee Calcarenite extends from the Sutakama River east to the Vaghanambo River, the Valasi Limestone between the Vaghanambo and Kaoka Bay.

The topography, particularly of the larger areas of the Mbohehe Limestone, tends to be karstic, with the development of sink-holes, steep river gorges and caverns (e.g. below Mbetilonga village).

The interfluves are characterised by 'numerous, very steepsided, rounded hills' (Allum, 1967) with occasional exposures of limestone, generally recrystallised and massive. The drainage tends to be deeply incised, and coarse with fairly sharp angles. East of the Lungga River the Mbetilonga Group forms a distinct escarpment facing southwards; the boundary with the underlying pre-Miocene rocks can be clearly delineated on the aerial photographs: in particular the northerly re-entrants into the Kolombolavu and Mongga valleys are very distinctive.

Reference localities for the various formations of the Mbetilonga Group are as follows:

Mbonehe Limestone	The upper Mbonehe River
Mbetilonga Limestone	The upper Tinahulu River (a section undisturbed by faulting has not been found)
Tina Calcarenite	The upper Tinahulu River
Lake Lee Calcarenite	The middle Mongga River, below Lee's Lake
Valasi Limestone	The lower Vaghanambo gorge

Diagnostic larger benthonic foraminifera are less common in the Mbonehe Limestone than in the other formations: details for the western and central areas are given in Coleman (1957; 1963). Algal, coral and, less commonly, molluscan remains may also be found. The gastropod *Conus (Chelyconus) loroisii* (Kiener) was discovered in the Hovu River section of the Lake Lee Calcarenite (identified by H. Ladd). The calcarenites of the eastern part of the Valasi Limestone outcrop are rich in planktonic and larger benthonic foraminifera, notably *Planorbulinella, Operculina complanata, Operculina venosa, Operculinella sp., Marginopora vertebralis, Alveolinella quoii* and *Amphistegina sp.* The alga *Halimeda* is common in the Valasi Limestone.

On the basis of the larger benthonic foraminifera, Coleman (1957 and 1960b) assigned an early Lower Miocene or e5 (approximately Aquitanian) age to the Mbetilonga Limestone and a Burdigalian (late Lower Miocene or f1) age to the Mbonehe Limestone, the Tina and Lake Lee Calcarenites. The microfossil assemblage of the Valasi Limestone is not diagnostic of any particular age-division, but indicates that the age is probably Pliocene, and certainly not older than Upper Miocene (Coleman, oral communication, 1969).

The Mbonehe Limestone rests unconformably on a 'basement of strong relief' (Pudsey-Dawson and Thompson, 1958). In the Mauvo Raha gorge of the Kolokumaha River the Lake Lee Calcarenite is nonconformable on an eroded surface of Mbirao Metabasics; the contact between the Valasi Limestone and the underlying Guadalcanal Gabbro is also visible in the Vaghanambo River. However, the nature of the contact is usually difficult to interpret owing to lack of exposure in the critical areas.

Locally the boundary may be tectonised: in the Tina River, 25 m below the Mbeambea-Mbicho confluence, schistose Mbirao Metabasics have been thrust in a north-easterly direction over thinly bedded flaggy calcarenites of the Mbetilonga Limestone formation.

Due to faulting, the height of the base of the Mbetilonga Group above sea level varies considerably.

In the Na Ndoli River, eastern Guadalcanal, a coarse breccia (Tpx) at least 50 m thick, comprising angular basalt fragments up to 3 cm in diameter, rests unconformably on bedded calcarenite which shows no signs of brecciation or shearing: it appears that an eroded surface of Valasi Limestone has been covered with basalt scree derived from a nearby palaeogeographic 'high' of 'Valasi Block' material.

The Tina, Lake Lee and Valasi formations are succeeded conformably and in part transitionally, by the fine-grained arenites of the Tangareso Beds and Mbokokimbo Formation. According to Pudsey-Dawson and Thompson (1958) the deposition of the Mbonehe Limestone was succeeded by a period of uplift and weathering, which was followed by the deposition of Pliocene tuffs and agglomerates.

The Valasi Limestone dips at about 10° to the north-east (Figure 13, sub-area 10B). However, the older formations have been gently warped about axes trending north-west. (Figure 13, sub-areas 9A and 10A); fuller structural details are given in Chapter 3. Comparable data for the Mbonehe Limestone are not available, but limited information to date shows a considerable variation in direction and amount of dip.

Thicknesses for the formations of the Mbetilonga Group, with minima estimated from vertical sections, are as follows:

Mbonehe Limestone	100 m
Mbetilonga Limestone	200–400 m
Tina Calcarenite	100–300 m
Lake Lee Calcarenite	100–350 m
Valasi Limestone	A maximum of 1000 m (between the Valasi and Na Ndoli rivers)

The Mbetilonga Group is composed essentially of carbonate rocks with a variety of texture, grain size and visible organic content: there is characteristically a high admixture of terrigenous clasts.

Biostromal limestone, typically a massive grey-weathering off-white pure carbonate rock, is usually recrystallised, at least in part. Coralgal and foraminiferal remains can often be identified, but molluscan fragments are rarely preserved. Such rock probably forms up to 70 per cent of the Mbonehe Limestone and is well represented in the upper half of the Mbetilonga Limestone and the basal beds of the Lake Lee Calcarenite. The western part of the Valasi Limestone outcrop is made up of a coralgal reef facies, well exposed around Isipeni on the Rere River.

Foraminiferal biocalcarenites or microcoquinoids, usually thick-bedded, may be intercalated with or grade laterally into the more massive recrystallised limestones. They may carry up to 30 per cent of angular terrigenous clasts, usually crystal fragments of fresh pyroxene and plagioclase feldspar, also lithic fragments of basic schist, basalt and basaltic andesite. There is a variable proportion of microcrystalline carbonate matrix; much of the material could be described as impure biomicrite in the system of Folk (1959, 1962). Such rocks are well represented throughout the Lake Lee

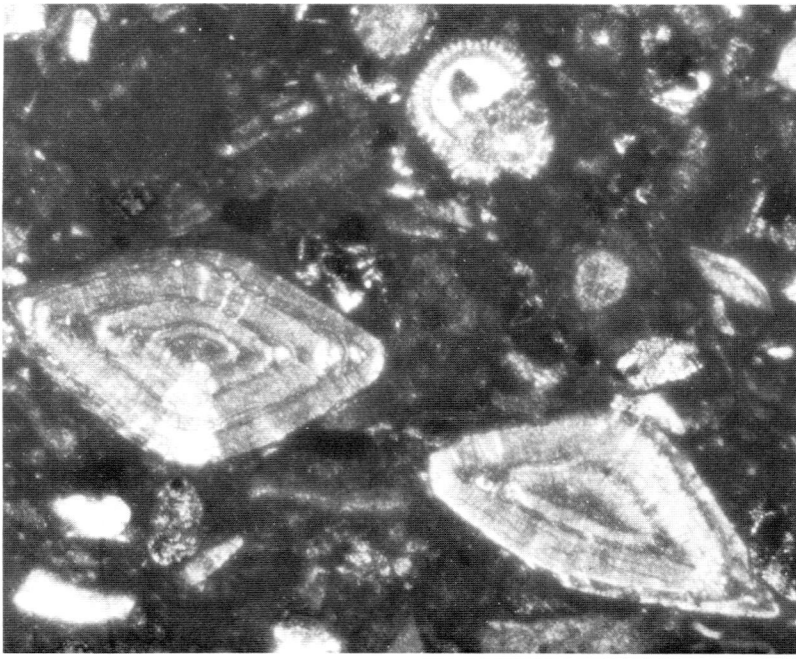

Plate 11 Photomicrograph of biomicrite with lithic fragments, Valasi Limestone, Na Koilo tributary of Aola River, east-central Guadalcanal. Benthonic and planktonic foraminifera, with angular lithic fragments of altered basalt in a matrix of fine-grained brownish carbonate

Calcarenite and in the eastern part of the Valasi Limestone outcrop (Plate 11). The Tina Calcarenite is predominantly a sequence of well-bedded flaggy calcarenites with a high terrigenous admixture: commonly they contain planktonic foraminifera, occasionally carbonised wood fragments.

In the Mbetilonga and Valasi Limestones the terrigenous content generally increases downsection, and the proportion of lithic fragments may exceed 50 per cent, that is the allowable limit for non-carbonate detritus in calcarenites according to Pettijohn (1957) and Carozzi (1960). A lens of such coarse material, the Tevua Conglomerate (Tmc) has been mapped as a subsidiary member of the upper part of the Lake Lee Calcarenite.

In the Sutakama Valley the basal 100 m of the Mbetilonga Limestone is highly conglomeratic: most of the clasts are demonstrably of very local origin, including a high proportion of serpentinite and pyroxenite derived from the nearby Suta Ultrabasics.

Calcisiltites or impure fine-grained calcareous arenites form a sequence at the top of the Lake Lee Calcarenite transitional to the Mbokokimbo Formation.

Lenses of calcirudite and coquina nowhere bulk large, but feature in the Vaghanambo and Kolonggao sections of the Valasi Limestone. Dolomites have not been observed.

Structures suggestive of penecontemporaneous brecciation and bioturbation have been observed in the biostromal facies of the Mbetilonga Limestone. The calcisiltites of the upper part of the Lake Lee Calcarenite feature some interesting structures, notably 'ball and pillow' effects (Smith, 1916), wave ripples and discordant flammate contacts.

For the bulk of the contaminated carbonate rocks in the Mbetilonga Group the energy index (Plumley and others, 1962) must be regarded as high. However, the cleaner biostromal limestones must have accumulated in quieter back-reef conditions or as fringing reefs, sheltered from liability to contamination by rubble from a steep unstable coastline. The features of the fine-grained calcarenite sequences indicate reworking by turbidity currents, probably in a fairly shallow sheltered embayment or interreef environment. The presence of *Halimeda* in the Valasi Limestone suggests a water depth of less than 60 m (McKee and others, 1959).

The Tangareso Beds

The Tangaraisu Marls of Coleman (1960b) or Tangaraisu Shale (Coleman and Day, 1965) can be mapped as a distinct formation, called here informally the Tangareso Beds, between the Mbetilonga Basin and the Sutakama Valley. Continuity of the outcrop is broken between the Tina and Toni rivers, and on the north side of the Sutakama.

The Tangareso Beds form a narrow belt of steep ground, overshadowed by the escarpment of the overlying Toni Formation. The profiles are more rounded, gentler in aspect than in other parts of the Upper Foothill Zone; river sections are mostly well graded and easily accessible.

All sections are disturbed to some extent by faulting: Coleman and McTavish (1964) refer to the upper part of the Tangareso River, where the formation was first described.

Planktonic foraminifera predominate in most thin sections: Coleman (1960b; 1963) has also recorded a large number of benthonic foraminifera. The lower part of the Tangareso Beds contains large numbers of *Operculina complanata*; in particular the basal beds in the Sutakama Valley exposures are crammed with individuals of this species.

According to Coleman and Day (1965) the foraminifera suggest a Middle Miocene age for the Tangareso Beds; they are overlain directly by the Charikangge Grit, which is probably Vindobonian (Coleman and McTavish, 1964).

The Tangareso Beds overlie the Tina Calcarenite with some measure of unconformity, and rest directly on Guadalcanal Gabbro in the Tinahulu headwaters and on ultrabasics in the Sutakama River. The Tangareso Beds were warped together with the underlying Mbetilonga Group, and partially eroded before the deposition of the Toni Conglomerates. Thus the Tangareso Beds fail to outcrop in the Tausoro Valley between the Tina Calcarenite and the Toni Formation. The Tangareso Beds tend to dip at fairly high angles to the north-east, but there is a wide range in amount and direction.

The maximum thickness of the Tangareso Beds is about 800 m in the area west of the Tina River; east of the upper Chovohio River the unit becomes progressively thinner and less persistent.

The dominant lithology is that of a rather poorly-sorted fine-grained arenite; the grade varies from that of fine silt to medium-grained sand on the revised Udden-Wentworth scale (Wentworth, 1933). Weathered surfaces are dull grey, occasionally brownish-grey, but fresh samples are usually bluish-grey, speckled with white calcareous fragments which mostly prove to be the tests of planktonic foraminifera. Broken molluscan fragments are occasionally found in the coarser sandy grades. In most thin sections the terrigenous fragments are highly angular: they include fresh crystal fragments of plagioclase, clinopyroxene, epidote and chloritised hornblende (Plate 12). Some laminae are microcoquinoid, made up almost entirely of planktonic foraminifera.

In the Tinahulu River section the content of the microcrystalline calcite tends to increase towards the base of the Tangareso Beds, indicating a transition to the underlying Tina Calcarenite; here the siltstones alternate with flaggy grey calcarenite containing macerated plant debris and clasts of Lower Miocene biomicrite. In the Sutakama Valley there are sporadic minor lenses of coralgal limestone and conglomerate.

Stratification is usually well formed (thickly laminated to medium-bedded in the scheme of McKee and Weir, 1953).

Small-scale cross-bedding occurs towards the base of the formation in calcarenite beds in the Chovohio River; the inclination of the laminae indicates derivation from the north-east. Higher in the same section asymmetrical deformed load casts are consistently overturned towards the west or south-west. Penecontemporaneous brecciation and patterns of desiccation cracks have also been observed.

The foraminiferal content of the Tangareso Beds is largely planktonic, indicating access to open ocean, but the bulk of the material is terrigenous. Dredging operations in the bottom sediments of the channel between Santa Isabel and San Jorge Islands revealed a floor of grey foraminiferal silt (Maranzana, 1968; oral communication); when dry and coherent, this was lithologically identical to the Tangareso arenites. A comparable environment, that of a bay or inter-insular channel connected directly to the ocean, into which rivers discharged silt derived from a basic volcanic terrain suggests itself as a palaeogeographic setting for the Tangareso Beds. The directional structures are also in accord with such a concept.

The Mbokokimbo Formation

The term Mbokokimbo Formation is proposed for a sequence of fine-grained arenites of Upper Miocene and Pliocene age which outcrop over an area of about 550 km² in east-central Guadalcanal; this name is an extension of the term Mbokokimbo Beds (Coleman, 1965a). The formation has complex intertonguing relationships with its neighbours, and could be described as a lithosome in the sense of Wheeler and Mallory (1956). The lithosome embraces the following subsidiary units:

Lower Mbokokimbo Member   Tmm
Upper Mbokokimbo Member   Tpm
Mbalanga Shale Member   Tps
Kolohaisava Mudstone Member   Tpk

The gentle rounded profiles contrast dramatically with the topography founded on the underlying rocks. Between the Manuhoho River and Lee's Lake the escarpment of Lake Lee Calcarenite is surmounted by a sheet of Mbokokimbo Beds inclined at a low angle to the north-west. Regional dips may readily be interpreted from the aerial photographs, as indicated on the geological map. The main rivers have a tendency to parallelism, but the tributary systems are distinctly dendritic; some of the larger tributaries, such as the Kolovagharindi and Humbaro rivers in the Mongga system, have incised deep gorges into the sedimentary pile. Photogeologically there is a clear contrast between the Mbokokimbo Formation and the interfingering conglomerates of the Lower Mongga–Aola area. However, Allum (1967) remarked on the variability of the photogeological appearance of this area: in part the Mbokokimbo–Mberande interfluves are obscured by relics of a Pleistocene erosional surface, making the delineation of boundaries difficult.

The Lower Mbokokimbo beds are exposed continuously in the Kolombolavu River from a point 1.6 km north of Haimbau village as far as the confluence with the Manuhoho River.

The Upper Mbokokimbo beds can be studied in the Mongga River below the Kolovagharindi confluence; as sections in the lower part of the valley are obscured by alluvium, supplementary information has to be obtained from the tributaries. The Mbalanga Shale is well exposed in the upper half of the Mbalanga River and its tributary, the Lame.

The Vaghanambo River and its lower tributaries provide the most complete sections of the Kolohaisava Mudstone Member.

As in the Tangareso Beds, planktonic foraminifera predominate throughout. However, in the coarser arenites towards the base of the succession and near the boundaries with the interfingering conglomerates, larger benthonic forms are found together with the planktonic species. Benthonic foraminifera derived from the Lake Lee Calcarenite are often found in lithic fragments near the base of the sequence.

The following forms have been identified from the middle and upper parts of the Lower Mbokokimbo Member: *Calcarina spengleri* (Gmelin), *Marginopora vertebralis*, *Amphistegina lessonii*, *Elphidium spp.*, *Operculina sp.* and the alga *Halimeda*.

Coleman (1960b; 1963) recorded the following foraminifera from the Mbokokimbo River section of the upper beds:
Alveolinella quoii (d'Orbigny)
Amphistegina lessonii
Cycloclypeus carpenteri
Globorotalia sp.
Globigerina spp.
Heterostegina sp. cf. *orbicularis* (d'Orbigny)
Operculinella spp.
Pulleniatina obliquiloculata
Sphaeroidinella dehiscens
Orbulina universa.

Shallow-water forms, including *Gypsina globosa* and *Amphistegina lessonii*, also abundant coralline algae, have been recorded from an intercalated calcirudite bed near the base of the Kolohaisava Mudstone in the Purasakusu River.

Bivalves occur in small 'nests' in the coarser sandstones of the transition beds which immediately underlie the conglomerates of the Paripao Tongue and Vatumbulu Beds of the Toni Formation.

Plate 12 Silty biomicrite, Tangareso Beds, Tinahulu River, north-central Guadalcanal. Plane polarised light. The rock is composed largely of planktonic foraminifera, with some fine-grained terrigenous arenite, mostly opaque

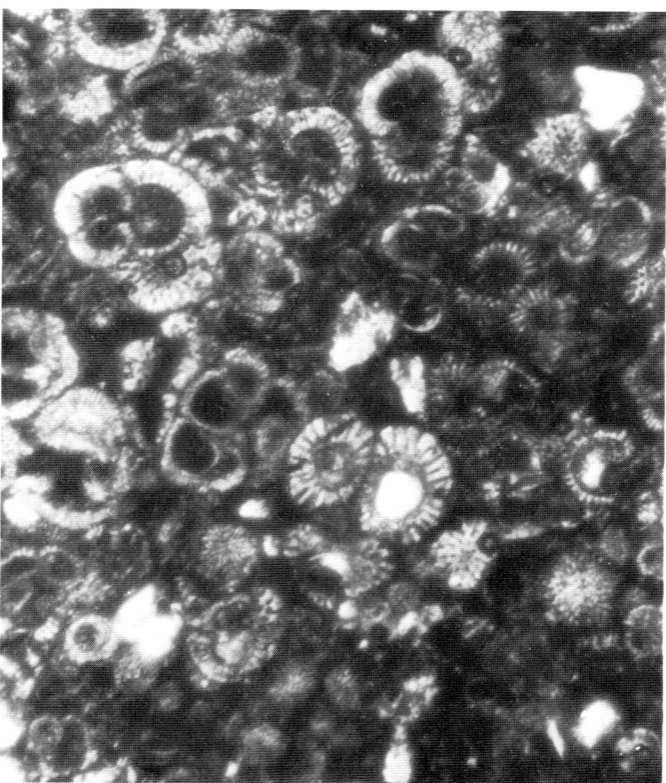

Carbonised wood fragments, leaf impressions and other plant remains are frequently found in the coarser silts and sandstones. Worm tracks have been observed in siltstones in the upper part of the Kolombolavu Gorge.

The larger benthonic foraminifera, and the general stratigraphic position, indicate a Pliocene age for the bulk of the Mbokokimbo Formation; however the extreme older limit for the basal beds is regarded as Upper Miocene. On the basis of planktonic foraminifera included in four samples from the Mbokokimbo River, McTavish (1968) concluded that the 'Mberande Beds' were at least in part Upper Miocene: these samples were in fact collected from the middle part of the Mbokokimbo Formation as mapped by the writer. The Lower Mbokokimbo Member is tentatively designated 'Tmm', although delineation of the Miocene–Pliocene boundary must await more detailed palaeontological studies. It is also probable that the youngest beds of the Upper Mbokokimbo Member are of Pleistocene age.

The shape of the Mbokokimbo Lithosome and the presumed relationships with adjacent formations are indicated in the composite stratigraphic section (Figure 4).

Intraformational disconformities occur near the base of the Lower Mbokokimbo beds and in the upper part of the Lake Lee Calcarenite in the Mavugho Gorge section of the Mongga River; it seems probable that major disconformities may occur at or near the base of the Mbokokimbo Formation, accounting for the non-appearance of the Middle Miocene.

The Lower Mbokokimbo beds are overlain in the west by conglomerates of the Mberande Tongue and to the east by the Vatumbulu Beds; in the middle reaches of the Mbokokimbo, Mongga and Aola rivers there is no change in lithology up-section through the middle of the lithosome.

In a central belt about 3 km wide which trends east-south-east between the upper Mbalanga River and the Aola headwaters there is considerable variation in azimuth and amount of dip, also evidence for penecontemporaneous faulting and slumping; the base of this zone provides an arbitrary upper boundary for the lower member of the Mbokokimbo Formation.

Stratification is usually well-defined and may be measured with ease and accuracy. Figure 13 (sub-area 10c) shows the attitude of bedding in the Lower Mbokokimbo Member; although the dips tend to cluster around 10° to the north-west, there is a distinct girdle conforming with the attitude of the underlying Lake Lee Calcarenite. Sub-area 10E indicates bedding attitudes in the Mbalanga Shale and underlying conglomerates of the Mberande Tongue. Although the dips cluster around a maximum of 5° to 10° to the north or north-north-east, there is a wide scatter due to localised penecontemporaneous deformation, and also, in all probability, to post-depositional compaction on an irregular sea-floor.

The Upper Mbokokimbo beds and Kolohaisava Mudstone dip at 5° to 20° to the north-east or north-north-east (Figure 13, sub-areas 10D and 10F). The stratification broadly conforms with the attitude of the underlying Valasi Limestone and the overlying Vatumbulu Beds. To the west the Upper Mbokokimbo Member thins rapidly, conformably overlying the conglomerates of the Paripao Tongue; rela-

tionships with the Honiara Beds are obscured by alluvium and exposure failure.

The Lower Mbokokimbo Member is 2000 m thick in the Kolombolavu River section, but thins rapidly westward to 200 m in the Kolokumaha River; probably the sequence does not connect directly with the Tangareso Beds, and there were two separate basins of deposition.

The Upper Mbokokimbo Member may be thicker than 2000 m, but the upper limits are difficult to define owing to the paucity of exposures in the area between Reko and Aola Bay.

The Mbalanga Shale Member is approximately 1000 m thick in the Mbalanga Valley, but thins rapidly westward. The Kolohaisava Mudstone is 2000 m thick in the Vaghanambo Valley, but thins to about 300 m east of the Valasi River.

The Mbokokimbo Formation is a monotonous sequence of poorly sorted arenites with planktonic foraminifera, similar lithologically to the Tangareso Beds. The grade may vary from clay to coarse sand, but fine-grained sandstones and siltstones predominate. The terrigenous fragments are usually subangular to highly angular; they may include crystal fragments of plagioclase (often quite fresh), clinopyroxene, rare hypersthene, vein quartz, leucoxene, epidote, chlorite and magnetite, also lithic fragments of feldsparphyric and hyalo-ophitic basalt, devitrified basaltic glass, saussuritised dolerite and chlorite-actinolite-schist. Detrital hornblende is common only in the Upper Mbokokimbo beds.

There are minor local variations within the arenites; dense black sandstones represent heavy concentrations of detrital iron ore, whilst pale blue calcareous mudstones or marls, weathering to a creamy off-white, are composed largely of planktonic foraminifera set in a sparse matrix of terrigenous lutite. Some features observed in thin sections of the finer arenites, such as the occurrence of siderite spherulites, pyrite and recrystallised carbonate are attributable to diagenetic processes.

In the middle Kolombolavu River there are occasional beds of coarse pebbly sandstone, fairly well-sorted with rounded blade-shaped clasts of basic schists.

In general the fraction of microcrystalline carbonate increases down-section, that is towards the top of the Lake Lee Calcarenite, and is particularly evident in the basal 300 m of the Lower Mbokokimbo beds; beds of coralgal rubble and coquina occur intercalated with the siltstones of the upper Aola River section. Minor lenses of concretionary white limestone, algal pisolite and calcirudite are found in different parts of the sequence.

Sporadic conglomeratic layers may occur throughout the sequence, either as paraconglomeratic or 'tilloid' lenses in 'cut and fill' structures or as thin orthoconglomeratic beds intercalated with coarse arenites; they are particularly common in the lower part of the Mbalanga Shale Member and in the zones of transition to the adjacent conglomeratic 'tongues'. The upper boundary of the Mbalanga Shale is not clear-cut: most river sections indicate a transition to the conglomerates of the Paripao Tongue, comprising a sequence of rapidly alternating conglomerates, sandstones and siltstones which may exceed 50 m in thickness. The included sandstones are better-sorted than the silty arenites of the central basin; for example, in the Sarivuro tributary of the Mbalanga clean pebbly sandstones alternate with well-sorted black ilmenite sands.

Phenoclasts in the conglomerates of the Lower Mbokokimbo Member include, in addition to the lithic fragments detailed for the arenites, uralite-gabbro and serpentinised harzburgite. In the Mbalanga Shale intercalated rudites also feature fragments of feldsparphyric basaltic andesite, hornblende-andesite and silicified volcanic breccia.

The basal 300 m of the Mbokokimbo Formation, exposed in the Mavugho Gorge of the Mongga River, exhibit a variety of interesting sedimentary structures; the section shows a transition from the top of the Lake Lee Calcarenite, in which grey muddy siltstones alternate rapidly with coarser, rather poorly sorted arenites and more massive calcisiltites. The siltstones, rich in carbonised wood fragments, show extensive wave rippling, in part asymmetrical; the crests of the ripple waves trend consistently east–west.

Large scale cross-bedding occurs in some of the coarser arenite beds towards the top of the Mavugho Gorge section; the planes of cross-stratification are generally inclined steeply towards the north-west.

In the Mongga River 1 km above the site of Mbumbunuhu village, the siltstones are diversified by intraformational disconformities with pockets of paraconglomerate in 'cut and fill' structures. Penecontemporaneous brecciation, load injection features and intrastratal slumping have been observed elsewhere in the sequence, notably in the Kolombolavu River above the Na Asa Gorge. In the latter section, sporadic concretionary structures occur in the darker siltstones; superficially these appear as spheroidal or ellipsoidal nodules with a rusty outer skin, ranging from 2 to 5 cm in diameter. The core is usually of homogeneous fine-grained calcarenite covered with a thin pyritised sandy envelope; possibly an original nucleus of organic material decomposed under anaerobic conditions to form a skin enriched in authigenic pyrite.

Desiccation cracks with hexagonal patterns have been found on bedding surfaces in the Lower Mbokokimbo Member and in the Kolohaisava Mudstone.

In the middle part of the Mbokokimbo Lithosome there are frequent paraconglomeratic intercalations: lenses of tilloid material fill steep channels cut into a substratum of stratified siltstone. Minor disharmonic folds, penecontemporaneous faults, slumping and bedding slips attest to increased instability in the depositional environment.

Dislocated slabs (olistoclasts?) of fine-grained laminated dark-grey siltstone may be enclosed in rather massive light-grey silty calcilutite. The laminations within the slabs are broadly parallel to the bedding in the host sediment; the structure is not chaotic as in the classic argille scagliose, but indicates that the siltstones were fractured shortly after consolidation and glided, probably on a semi-coherent layer of mud, to be incorporated within still unconsolidated downslope sediment.

The Mbalanga Shale shows a variety of structures, especially near the boundary with the overlying conglomerates of the Paripao Tongue. Wave ripple-bedding showing distinctive 'cusps', also medium-scale cross-stratification and ripple cross-lamination (McKee, 1938, p. 80) are common

features; large-scale cross-bedding may occur in units up to 3 m in thickness. Crude grading, localised penecontemporaneous breccias and minor intraformational disconformities have also been recorded from the upper part of the Mbalanga River section.

In the Njarikusu tributary of the Mongga, small-scale cross-stratification occurs in thin sandy beds intercalated with conglomerates: the planes of cross-bedding are inclined towards the south-east. Wave ripple-bedding is found in the same river section, the axes of the ripples trending consistently north-north-east.

Reference to the key to environmental interpretation formulated by Conybeare and Crook (1968, pp. 58–60) suggests that the sedimentary structures of the Mbokokimbo Formation are characteristic of fluviatile and neritic environments. The abundance of planktonic foraminifera indicates access to the open sea; the lithology and structures suggest deposition by continuous turbid flow, punctuated by occasional phases of marked instability, which could be attributed to seismic disturbances.

It is suggested that the Mbokokimbo Lithosome represents the site of a shallow gulf or estuarine complex, the Mbokokimbo Basin, which was maintained from the Upper Miocene to the Pleistocene, receiving large quantities of silt from rivers draining a basic volcanic terrain. The interfingering conglomerates indicate periods of shoaling and restriction of the marine embayment, due to uplift or increase in the sediment supply from an unstable source area.

Anomalous dips in the middle part of the sequence may have resulted from differential compaction over an irregular basement surface; the disturbed zone does, however, closely parallel the buried boundary of a geophysical unit (Figure 24). Pliocene movement along a buried fault line, concomitant with gentle warping of the Miocene sediments, could account for the anomalous zone.

The Toni Formation

The Toni Formation is a sequence of volcaniclastic rudites and arenites with subsidiary pyroclastics, extrusives, and biogenic limestones, ranging in age from Middle Miocene to Upper Pliocene. The formation extends across north-central Guadalcanal in the foothills zone between the Lungga and Mbokokimbo Rivers. Its name may be regarded as a development of the terms Toni Tuffs (Coleman, 1960b) and Toni Beds (Coleman and Day, 1965).

Considerable horizontal variations in lithofacies give, in detail, a very complex outcrop pattern (see Geological map of Guadalcanal). The main subdivisions have been mapped as follows:

1　The Charikangge Grit (the Charikange Beds of Coleman and McTavish, 1964).
2　The Haviha Sandstone.
3　The Lower and Upper Toni Members*.
4　The Gold Ridge Volcanics.
5　The Mbetivatu Sandstone.
6　The Kombusoe Limestones.

*Together with two easterly prolongations, the Mberande Tongue and the Paripao Tongue, these units form an irregular mass of conglomerate, the Toni Lithosome (see Geological map of Guadalcanal).

Boundaries between the various members have been mapped to show those changes in lithology which affect a significantly large area on the 1:50 000 scale. These changes are not everywhere abrupt; subtle variations could be mapped in different ways according to the emphasis of the field interpretation.

The country is rugged, deeply dissected by a dense drainage network, and on aerial photographs appears lighter in tone than the adjacent areas of igneous and arenaceous rocks. Where the Toni conglomerates overlie Middle Miocene siltstones, they form a distinct escarpment presenting a steep face to the south. The rivers have cut narrow ravines, choked with large boulders of resistant volcanic rocks which were originally phenoclasts in the surrounding rudites. The Mbetivatu Sandstone tends to form lower, less steeply dissected ground than the conglomerates; however the boundaries between various members are not readily delineated on aerial photographs: the Gold Ridge Volcanics outcrop is in no way distinct from that of the Toni conglomerates.

The Charikangge Grit outcrops in the section of the Tangareso River which extends for approximately 1 km above its confluence with the Charikangge. The Haviha Sandstone is exposed only in the headwaters of the Charihoro tributary of the Kolokumaha River.

The Toni tributary of the Matepono River provides a good reference section for the Lower Toni Member, but exposures are nearly continuous throughout both the Lower and Upper Toni Members in the Tinahulu River (extensively faulted) also in the middle Chovohio River and its tributary, the Tausoro.

The Gold Ridge Volcanic Member is well exposed in the upper Chovohio River for a distance of 5 km above its confluence with the Charivunga.

The Mbetivatu Sandstone is best exposed in the Mbetivatu tributary of the Tangareso River (P. C. Wright, oral communication, 1965).

The Kombusoe Limestones are biostromal lenses of small extent, exposed sporadically in the Chariturua, Mamasa and Charihau tributaries of the Chovohio River.

Coleman and McTavish (1964) list a large number of foraminifera from the Charikangge Grit; an association of larger benthonic and planktonic foraminifera of the 'Sphaeroidinellopsis seminulina fauna' was found, even within single samples. Derived Lower Miocene foraminifera are often found in biomicrite clasts in the Toni conglomerates; much of the sequence, however, is barren. *Calcarina sp.* has been found in the conglomerates of the Mberande Tongue (Mberande River) and shallow-water forms such as *Elphidium spp.* in the Paripao Tongue (Sotokiki River). Shadowy remnants of planktonic foraminifera have been detected in thin sections of tuffaceous arenites intercalated with the Gold Ridge Volcanics.

In the Mbetivatu Sandstone, a sample from the Chariturua River yielded *Operculina complanata*, *Operculina venosa*, *Amphistegina lessonii* and *Cycloclypeus sp.* cf. *indopacificus*. McTavish (1968) gives details of more planktonic foraminifera from the Tangareso River section of the Mbetivatu Sandstone. A shelly mudstone from the Charihau River contained numerous specimens of the freshwater snail *Sermyla* (H. Ladd, written communication, 1970), molluscan fragments, and much carbonaceous plant debris. The alga *Halimeda* is an important constituent of the Kombusoe Limestones.

Coleman and McTavish (1964) infer a probable Middle Miocene (Vindobonian) age for the Charikangge Grit. The foraminifera from the Mberande Tongue are no younger than Lower Pliocene; for the Paripao Tongue an Upper Pliocene age is indicated. Probably the Gold Ridge Volcanics range through the Upper Miocene and Lower Pliocene; in general, dating of the Toni Formation is hampered by the poverty of diagnostic short-range forms.

In the Tinahulu River, the conglomerates of the Lower Toni Member conformably overlie the Middle Miocene Tangareso Beds. However, the base of the Toni Formation successively overlaps onto the Mbetilonga Group, the Suta Ultrabasics and the Guadalcanal Gabbro. To the east the Toni and Mbokokimbo Lithosomes are involved in a complex interfingering relationship. To the west more mapping is required to clarify the details of transition to the Lungga Beds (Tpl). Enveloped by the Toni Lithosome, the Mbetivatu Sandstone may be mapped as a distinct unit between the Tangareso and Chovohio rivers: elsewhere its limits are ill-defined. The eastern and western boundaries of the main diamond-shaped outcrop of the Gold Ridge Volcanics are not exposed: it is presumed to be a lens with very rapid lateral transition to the surrounding Toni conglomerates.

In Figure 13 sub-areas 9C and 9D show the variation in the direction of dip, although the modal maximum is around 10° towards the north-north-east or north-east. The scatter is due to localised faulting, also to the initial irregularities in the bedding and the consequent difficulties of making representative measurements.

The Charikangge Grit forms a lens about 80 m thick, which is not mappable as a unit outside the Tangareso Valley; thus the name has very limited application. The Haviha Sandstone is a similar lens about 200 m thick.

The maximum thicknesses of the subdivisions of the Toni Lithosome are as follows:

Lower Toni Beds	600 m
Upper Toni Beds	300 m
Mberande Tongue	1000 m
Paripao Tongue	700 m

The Gold Ridge Volcanic Member has a maximum thickness of about 800 m. The thickness of the Mbetivatu Sandstone ranges from about 800 m in the Tangareso Valley to less than 200 m in the Chovohio. The Kombusoe Limestones nowhere exceed about 100 m in thickness.

Lithology

THE CHARIKANGGE GRIT AND HAVIHA SANDSTONE
The Charikangge Grit is a lens consisting predominantly of poorly bedded volcanic wackes with calcareous cement (Coleman and McTavish, 1964). The Haviha Sandstone is a sequence of grey-weathering poorly sorted flaggy calcareous arenites, in which the proportion of intercalated paraconglomerate increases up-section.

THE TONI LITHOSOME   The lithology of the Toni Lithosome is essentially rudaceous, the material being mostly of volcaniclastic origin. Orthoconglomerate (Pettijohn, 1957) predominates in those parts of the Lower Toni and Mberande Tongue members which directly overlie the older sediments and basement. Paraconglomerates are found in the Middle Mberande, Sutakama and Chovohio river sections, underlying and passing gradually into the Mbetivatu Sandstone and Mbalanga Shale. The Upper Toni Member and the Paripao Tongue are predominantly orthoconglomeratic. The clasts vary in size from coarse silt to about 30 cm, although isolated boulders may exceed 10 m in diameter: they may be subangular or well rounded, but a majority of phenoclasts would be classed as subrounded, while the silty fragments in the matrix are highly angular. The matrix is locally calcareous.

A survey of the petrography of phenoclasts has been made for 220 different stations throughout the Toni Lithosome. Some of the rock types can be matched from elsewhere in Guadalcanal: accordingly each type has been assigned a letter category to indicate possible provenance according to the following scheme:

A Pre-Miocene (Mbirao Group).
B Suta Volcanics.
C Mbetilonga Group and Tangareso Beds.
D Upper Miocene–Pliocene.

The percentages of component types for clasts of approximately cobble size, have been calculated as follows:

	Per cent	Category
Feldsparphyric basaltic andesites	29	B or D
Miscellaneous non-porphyritic basalts	15	A, B or D
Hornblende andesites	14	D
Basalts porphyritic in green pyroxene	8	B or D
Andesites porphyritic in hypersthene	7	D
Sheared gabbro	7	A
Recrystallised limestones	6	C or D
Ultrabasic rocks	4	A
Calcisiltites and calcilutites	3	C
Miscellaneous metabasic rocks	2	A
Fresh gabbro	2	A or D
Miscellaneous pyroclastics	2	D
Conglomerates	1	D

The inferred provenance of the clasts may be summarised as follows:

15 per cent Pre-Miocene
 9 per cent Lower and Middle Miocene sediments
24 per cent Pliocene volcanics and sediments
52 per cent Uncertain: in part resembling Suta Volcanics

Towards the base of the conglomerates, a large proportion of the phenoclasts has been derived from the substratum: thus ultrabasic fragments constitute as much as 80 per cent of the material in the basal 30 m overlying the ultrabasics in the Sutakama Valley.

Phenoclasts of hypersthene-andesite can only be matched with a small lava flow exposed in the Kova River; it may be inferred from the occurrence in the conglomerates that flows of hypersthene-andesite might have been more extensive in the Pliocene, but were rapidly eroded or buried under later sediment.

MBETIVATU SANDSTONE This is typically a crudely stratified brownish-yellow arenite, weathering to a rusty brown or grey colour; paraconglomeratic lenses are fairly frequent.

The arenites vary somewhat in their degree of sorting, but are composed essentially of angular fragments of basic volcanic detritus, with rarer lithic fragments of mudstone and siltstone. Lithologically they are comparable with the matrix of the conglomerates of the Toni Lithosome, and could be described as volcanic and lithic arenites (Williams and others, 1954). Carbonised wood is locally abundant. Crystal fragments of fresh twinned plagioclase and basaltic hornblende may together make up 70 per cent of the rock; some laminae are exceptionally rich in magnetic iron ore. Locally the matrix is calcareous, and the arenites grade into biomicrites and fine grey marls with occasional bands rich in molluscan remains.

In the intercalated paraconglomerates the proportion of hornblende-andesite phenoclasts was about 40 per cent for a survey of 63 stations: this is significantly higher than for the bulk of the Toni Lithosome.

KOMBUSOE LIMESTONES These are lenses of massive coralgal material, usually in part recrystallised, alternating with bedded calcarenite. In the Mamasa River gritty calcarenite fills channels in the fretted surfaces of underlying coralline rock; with increasing proportions of terrigenous clasts, the calcarenites grade into the surrounding volcanic arenites of the Mbetivatu Sandstone.

Sedimentary structures

The Charikangge Grit shows small-scale cross-stratification, minor slumps and other intraformational structures, but the greater part of the Toni Lithosome appears to be a structureless conglomeratic sludge. The stratification is poorly defined; occasionally a crude imbrication is developed. Cross-bedding is rarely observed.

In the Middle Sutakama and Mberande river sections, beds of poorly sorted arenite overlie highly irregular surfaces of paraconglomerate; orthoconglomerate also fills steeply scoured channels in silty sandstone. A few sedimentary dykes have been mapped in this area: they are more or less vertical bodies of ill-sorted arenite, not exceeding 2 m in width, which cut through poorly bedded paraconglomerate. These dykes occur immediately below the sandstones which herald the transition to the Mbalanga Shale. Presumably the paraconglomerates were sufficiently coherent to have rifted under tension while the overlying arenites were being deposited.

Petrography of the Gold Ridge Volcanics

The Chovohio River section above the Charivunga confluence is almost entirely in volcanic breccia in the sense of Bailey (Wentworth and Williams, 1932): the rocks were described as fragmental flow breccias by Grover (1958b). Eighty per cent of the fragments exceed 32 mm in diameter, and occasional fragments are larger than 20 cm. The blocks are very angular and are predominantly of altered feldsparphyric basaltic andesite and vesicular basalt. No accidental fragments have been recorded. The colour of the breccias varies attractively from blue to purple, grey, brown and orange; however, green is the commonest shade.

The volcanic breccias are generally well stratified and are intercalated with laminated lenses of volcanic arenite and lapilli-tuff. The matrix of the breccias is a purplish-grey coarse sandy tuff, composed largely of zoned turbid plagioclase crystals, angular fragments of basaltic glass, and iron ore, and (occasionally) basaltic hornblende and lithic fragments of hornblende-andesite.

Only one unbrecciated basaltic flow was located, less than 2 m in thickness: the matrix of this rock was found to be composed of an aggregate of glass shards rimmed with palagonite.

In the basal 100 m of the Chovohio section the matrix is calcareous, due to admixture of microcrystalline carbonate. In the northern part of the main outcrop, notably in the lower Charivunga River, the breccias have been hydrothermally altered; the exposed rocks are highly weathered and friable. Common features are silicification, pyritisation, kaolinisation and the formation of secondary calcite, chlorites and hydrated iron ore.

The Gold Ridge Volcanic Member has been distinguished from the surrounding Toni conglomerates on the basis of the angularity of the clasts and the overwhelming predominance of volcanic material. However in the Njarivila tributary of the Sutakama and in the Chariveha River up to 200 m of well bedded coarse sandy tuffs and lapilli-tuffs are intercalated with the Mbetivatu Sandstone; these rocks have been mapped as Gold Ridge Volcanics, although distinction from the surrounding volcanic arenites is rather uncertain in the field.

Mineralisation

Quartz and pyrite are widely disseminated in the northern part of the Gold Ridge Volcanic outcrop. Gold is associated with quartz veins and stringers; no payable lode has been found in situ, although approximately 28 kg of alluvial gold are panned annually by Solomon Islanders. Malachite and chalcopyrite have been found in floaters in the middle Chovohio.

Figure 21, sub-area 9B, shows the strike pattern for gold-quartz veins: the structural details are discussed in Chapter 4.

The Chovohio Igneous Phase

Minor occurrences of igneous rock, mapped as the Chovohio Igneous Phase, may be considered together with the Gold Ridge Volcanics:

A flow of porphyritic hypersthene-andesite 3.8 m thick is exposed in the Kova tributary of the Chovohio River.

Dyke-like bodies of volcanic breccia up to 3 m in width and trending west-north-west, intrude massive orthoconglomerate between Old Case and Bagarice in the middle Chovohio River.

A porphyrite dyke 150 m wide, trending north-east, intrudes conglomerates in the Toni River: spectacular zoned phenocrysts of clinopyroxene are set in a hemicrystalline matrix.

The depositional environment of the Toni Formation

Coleman and McTavish (1964) suggest that the Charikangge Grit was deposited in a tectonically unstable environment: slumping and turbid flow transported shallow-water sediment into deeper water and mixed foraminiferal faunas diagnostic of differing environments. The poorly sorted immature petromict conglomerates of the Toni Lithosome were probably deposited partly in shallow seas and partly subaerially, representing a period of intense alluviation deriving its energy from an active volcanicity of Suta type, which produced basaltic andesite or tholeiitic extrusives in the Oligocene–Lower Miocene and continued intermittently in north-central Guadalcanal into the Pliocene; but the original volcanic pile, eroded and covered by epiclastic debris, is still exposed to view only in the area of Gold Ridge. Analysis of the clastic components suggests that large volumes of hypersthene-andesite and, latterly, hornblende-andesites, were extruded during this phase.

Petrographic and ecological information indicates that the Mbetivatu Sandstone was deposited in an estuarine environment; clearly it represents a relatively quiet interlude in a sequence of high-energy alluviation. Chaotic paraconglomeratic facies may have been, in part, 'slopped forth by tsunami off-surge' (Coleman, 1970).

The Mbalisuna Gabbro

Exposures of the Upper Pliocene(?) Mbalisuna Gabbro are confined to about 1.7 km of the lower Sutakama River, above the old village site of Mbeghovila, and also in the adjacent Mbeghovila tributary.

Contacts with the surrounding conglomerates of the upper part of the Toni Lithosome are obscured by alluvium. However the gabbro tends to become finer towards the southern margins, which have been chilled to a glassy blue basalt over a distance of 4 m; also clasts of dolerite with schlieren, resembling parts of the Mbalisuna Gabbro, are found in the conglomerates of the Paripao Tongue. Accordingly it is inferred that the gabbro was emplaced as a high-level dyke or sill into the Toni Formation.

The rock is a fresh medium- to coarse-grained augite-gabbro, green and black in colour, but purplish-grey on weathered surfaces. The coarser facies are variable in aspect, with sporadic clusters of mafic xenoliths, pyroxene-rich schlieren 3 to 4 cm thick and flow-banding which locally resembles a gneissose foliation. The flow bands dip steeply to the west and strike north-north-east, parallel to the Mbalisuna Discontinuity (Figure 24). Flow folding has been observed in derived floaters, but not in situ.

An analysis from the Mbalisuna Gabbro is presented in Table 6. Chemically it is an alkali-gabbro with normative olivine and nepheline, similar to some of the more alkaline representatives of the Mbirao Group. The TiO_2 content is relatively high. Normative feldspar is basic labradorite.

The Vatumbulu Beds

The Vatumbulu Beds (Hackman, 1968c) are predominantly conglomeratic sediments which outcrop in the lower foothill zone between the Aola and Simiu rivers.

This zone forms a distinctive belt of country, finely dissected by a dense network of creeks which carry little surface water except after heavy rains. Their reference locality is the Ko Ravu tributary of the Kombito River where they reach their maximum thickness of about 700 m. They dip at low angles to the north or north-north-east (Figure 13, sub-area 10F) and overlie the Kolohaisava Mudstones conformably: seaward they are succeeded by sediments of Honiara Beds type, and are blanketed by the Recent alluvials of the narrow coastal plain.

The alga *Halimeda* and recrystallised coralline material have been identified in a lens of biostromal limestone in the Mberina tributary of the Aola River. Planktonic foraminifera are common in intercalated siltstones towards the base of the sequence.

The Vatumbulu Beds are assumed to be of Upper Pliocene–Pleistocene age by virtue of their stratigraphic position; they are predominantly conglomeratic and are lithologically similar to the Toni Lithosome. The basal 50 to 80 m, exposed in the head of the Ko Ravu stream, is made up of grey siltstone with subordinate paraconglomerate in beds up to 3 m thick. The remainder of the sequence is entirely orthoconglomeratic with occasional beds of buff-grey siltstone and mudstone, nowhere exceeding a few metres in thickness.

The aspect of the orthoconglomerate is uniform throughout the outcrop. The matrix is a somewhat incoherent grey sand or silt, grey-blue where fresh, and often slightly calcareous. The phenoclasts are commonly subrounded, of pebble and cobble grade; the fragments are predominantly of altered metabasic rocks and purple basalt (typical of the Mbirao Volcanics of the Valasi Block); rare clasts of hornblende-andesite are found towards the western end of the outcrop.

About 30 m of cavernous recrystallised white coralgal limestone are exposed in the upper gorge of the Mberina River; there are intercalations up to 1.8 m thick of massive calcisiltite containing sparse angular fragments of fresh basalt and purple gabbro.

Wave ripple-marks, trending roughly north–south, have been observed in intercalations of interbedded siltstone and marl in the Mbuha River.

Like the Toni conglomerates, the Vatumbulu Beds are considered to be of paralic origin (predominantly alluvial and shallow neritic). However, most of the detrital material was derived from the Mbirao Group, rather than from contemporaneous volcanics as is the case with the Toni Formation.

PLIOCENE SEDIMENTS IN WESTERN GUADALCANAL

The equivalent of the Toni Formation in western Guadalcanal is the Lungga Beds (Wright, 1968b).

Essentially the Lungga Beds (Tpl) comprise a succession of arenites and wackes derived primarily from volcanic sources, with subsidiary conglomerates, mudstones and andesitic lava flows. Local facies changes may be very rapid.

Coleman (1957) gave the general name Lungga Volcanics to a series of pyroclastics and volcanic rocks which outcrop over about three-quarters of western Guadalcanal.

Pudsey-Dawson and Thompson (1958) retained the name, describing it as a 'Sedimentary Volcanic Group'. Their detailed description shows that there are many features in common with the Toni Formation: '. . . coarse thick and irregular beds of agglomerate developed around the gradually rising andesite cones; while away from the centre of vulcanicity thick beds of tuff were deposited.' An interesting parallel with the Toni Formation is the increase in the contribution of hornblendic andesite towards the top of the succession, either as hornblende fragments within the arenites, or as lithic fragments in the rudites.

Pillowed flows and water-laid tuffs are regarded as subsidiary to subaerial volcanics, but clearly there are rapid local changes in facies, with evidence of marine incursions.

Wright (1968b) extended the mapping into the central basin of the Lungga (Mbetikama) River; he preferred the term 'Lungga Beds', as he considered that the majority of the succession was of greywacke and subgreywacke-type sediments, very similar to the lithic and volcanic wackes of the Miocene of the Aure Trough in Papua (Edwards, 1950). Wright mapped minor hypabyssal and volcanic phases, but considered that subdivision on the lines of Coleman's earlier work was impracticable owing to rapid facies changes.

Coleman (1965a) considered that the Lungga Volcanics interfinger with the Mt Austen Beds (regarded here as part of the Toni Formation), and that both are probably of Pliocene age.

In his photogeological interpretation of western Guadalcanal Allum (1967) remarked on the similarity of appearance of his V1 volcanics and S2 sediments on air photographs. The V1 volcanics are in fact largely the Gallego Volcanics, but include in part older (Oligocene–Lower Miocene?) andesites. The S2 sediments are, in the west, the equivalent of the Lungga Beds of Wright, 1968b).

Thin sections of lithic arenites and lithic wackes from the Lungga Basin occasionally contain planktonic foraminifera; the proportions of lithic or crystal fragments vary considerably, and the matrix may be of microcrystalline calcite or terrigenous silt. The variation in composition and texture is greater than in the Kavo Greywacke Beds: in particular there is a variety of lithic fragments including feldsparphyric andesite, vesicular basalt, hornblende-andesite, hydrothermally altered andesite, and fragments of reworked sediments. They vary in grade from coarse sand down to fine silt or clay, and are generally poorly sorted. Lithologically these arenites are closely comparable with the Mbetivatu Sandstone.

The Lungga Beds show the same facies variations as the Toni Formation, with volcanics, pyroclastics, epiclastic sediments predominantly derived from volcanic material and subsidiary biostromal limestones. The inferred overall time range—Middle Miocene to Upper Pliocene–Pleistocene—is the same in both cases. In western Guadalcanal the Kavo Greywacke Beds pass up-section into the volcanic wackes and lithic arenites of the Lungga and Itna Basins. (G. W. Hughes, 1975, oral communication).

Rapid variation in facies, complicated by the 'blurring' of sedimentary environments due to active tectonism during accumulation (Krumbein and Sloss, 1963, p. 428) has given rise to a very complex picture. This can be unravelled only by the systematic mapping of lithological changes in detail over the whole of western Guadalcanal. Until this has been completed, the name 'Lungga Beds' remains as a satisfactory non-committal blanket term for these Middle Miocene–Upper Pliocene? sediments and volcanics, which are broadly equivalent to the Toni Formation in western Guadalcanal.

The Umasani and Mataniko tuffs

Between the Lungga and Umasani rivers, on the northern flanks of the Poha–Mbonehe ridge, much of the Pliocene sequence appears to be missing or very condensed (Figure 4, composite stratigraphic section).

Coleman (1957) described up to 10 m of pyroclastic sediment, best exposed in the Umasani and Mavo rivers, which overlie the Mbonehe (Bonegi) Limestone disconformably. To this he gave the name Umasani Tuff. Agglomeratic at the base, it is predominantly a calcareous tuffaceous arenite with sporadic foraminifera indicating a possible Upper Pliocene age.

The Mataniko (Metanakau) Tuffs, named by Coleman (1957) after a single traverse in a tributary of the upper Mataniko, were included under the term Lungga Volcanics by Pudsey-Dawson and Thompson (1958). At least 100 m of stratified tuffaceous arenite contain a different, possibly older, foraminiferal assemblage than that of the Umasani Tuffs, but nevertheless suggestive of a Pliocene age.

Until further attention is given to the stratigraphic subdivision of western Guadalcanal, these two units may be regarded provisionally as members of the Lungga Beds 'lithosome'.

The Mt Austen Beds and the Mberande (Berande) Beds

Coleman (1960b) applied the name Mt Austen Beds to 'a largely undifferentiated mass of sediments which overlie the Miocene sequence'. Typical sections were provided by the streams on Mt Austen, south of Honiara, and the middle reaches of the Tenaru River. They included the Tenaru Conglomerates (Tpt), now mapped as part of the Toni Lithosome, but not the Toni Tuffs; the Tenaru Conglomerates were considered to merge gradually into the overlying sequence of tuffs, grits and conglomerates of volcanic origin with subsidiary biostromal limestones.

In his photogeological interpretation Allum (1967) reserved this name for a very small fault-bounded area on Mt Austen itself (Tpa). He considered that the Mt Austen Beds appeared more resistant to erosion than the adjacent Honiara Beds, but it was doubtful whether they should be considered as a separate unit.

In the course of mapping it has been found that Coleman's Mt Austen Beds correspond to the Tenaru Conglomerates (in part), the Mbetivatu Sandstones, and the Honiara Beds (in part).

The Berande Beds were described by Coleman (1960b) as a sequence at least 1000 m thick overlying the Lake Lee Calcarenite, starting with white biostromal limestone with intercalated layers of dark blue and dark grey tuffs and tuffaceous shales, and grading into coarser pyroclastics towards the top.

Coleman and Day (1965) applied the name Berande Beds to the eastern extension of the Mt Austen Beds, and renamed the lower part of the succession the Bokokimbo Beds. The latter corresponds broadly with the writer's Mbokokimbo Formation, but the Berande Beds as remapped include parts of the Toni Formation, the Honiara Beds and the upper part of the Mbokokimbo Formation.

Accordingly, to avoid confusion it has been decided to abandon the names Mt Austen Beds and Berande Beds.

PLIOCENE–QUATERNARY IGNEOUS ACTIVITY

Igneous activity reached a climax in western Guadalcanal towards the end of the Pliocene; volcanic activity was concentrated in, although not confined to, the north-west corner of the island. Both volcanic and plutonic manifestations are described.

The Gallego Lavas

Pudsey-Dawson and Thompson (1958) separated the Gallego Lavas from the Lungga Volcanics of Coleman (1957). The term Gallego Volcanics is used in the legend for the geological map, as the outcrop delineated in western Guadalcanal includes a large volume of pyroclastic and derived sedimentary material.

The Gallego Lavas occupy a large area of north-west Guadalcanal; elsewhere only about 4 km² of hornblende-andesite of Gallego type have been mapped at Cape Henslow, the south-eastern extremity of the island.

In north-west Guadalcanal the Gallego Lavas form a group of steeply dissected cones with a well-developed radial drainage. The rugged features and light tone are distinctive on aerial photographs.

The Gallego Lavas were first described from the section at Mt Gallego, where they are 900 m thick and are well exposed in the tributaries on the northern side of the Umasani River.

A K:Ar determination (Dr N. J. Snelling, written communication) on a sample of hornblende-andesite from Aruliho (BSI 8198) gave an age of 6.392 ± 1.95 million years, i.e. Lower Pliocene. It follows that much of the volcanic sequence may be older than Pleistocene, although intermittent vulcanicity continues to the present day on nearby Savo Island.

Structural relationships are invariably obscured by exposure failure; in vertical sections Pudsey-Dawson and Thompson (1958) show the Gallego Lavas overlying the Lungga Beds; however, the results of work by R. A. Dennis (written communication, 1971) in the Aruliho area suggest that sedimentary and volcanic beds interdigitate. In the Cape Henslow area at least 200 m of Gallego-type andesites are succeeded disconformably by Pleistocene sediments, the Mbota Moli Beds. The main outcrop of the Gallego Lavas appears to be associated with the north-east-trending Savulei Fault Zone. Also at Cape Henslow the Gallego

Lavas are located at the convergence of three major fault lines.

Pudsey-Dawson and Thompson (1958) described the petrography of the Gallego Lavas in some detail. In andesite from the Cape Henslow occurrence the hornblende is a fresh basaltic variety ($2V \approx 60°$; $\gamma:Z = 8°–15°$). Some thin-sections show abundant accessory apatite. In hand-specimens lustrous black hornblende 0.5 to 7 mm in length, and white plagioclase, are porphyritic in a dense grey groundmass. Occasionally a rough parallelism of the phenocrysts is observed. Locally the andesites have been hydrothermally altered; the iron-rich amphibole has been replaced by pyrite and the feldspar kaolinised, the altered product weathering to a coherent yellow clay.

Three analyses of andesites from the Gallego suite are presented in Table 7, together with CIPW norms. The weight percentages of Al_2O_3 and $Na_2O + K_2O$ are distinctly higher than in the Lower Miocene Suta Volcanics; however the K_2O/Na_2O ratio is much the same for the two suites. On an 'F'MA diagram (Figure 29) the plots fall within the field of the Hypersthenic Series of Kuno (1959); petrologically they might be considered as belonging to the Calc-Alkali Rock Series; the mafic mineral evolution, with the development of hornblende and biotite phenocrysts, is typical of the calc-alkaline trend.

Considered together with the Koloula granodiorite, which is of comparable age to the Gallego Lavas, the silica/alkali ratios indicate a calcic trend with a Peacock Index possibly exceeding 60.

The Koloula Diorite

A dioritic body of stock-like dimensions outcrops over 24 km^2 in south-central Guadalcanal in the valleys of the Koloula and adjacent Ghoivara and Kolokoo rivers.

The diorite is not readily distinguished from the surrounding Suta Volcanics on aerial photographs; it is believed that the detailed outcrop picture should be more patchy than as shown on the 1:50 000 map, as the ridge summits between the Kolokoo and Koloula valleys are almost certainly capped with canopy material. Mapping the roof of the diorite is rendered difficult by the steepness of the terrain, the hydrothermal alteration and the screens of colluvial material.

The granodioritic core of the stock is best exposed in the Charilava tributary of the Koloula; the Ghecha River provides a section through a variety of facies, including augite-gabbro, quartz-biotite-diorite, hornblende-diorite and an altered mineralised zone.

Two samples of the Koloula Diorite were dated by the K:Ar method (Snelling, written communication, 1969). The results were as follows:

	BSI 6967	BSI 6970
Locality	Charilava River	Kolokuau River
Material analysed	Hornblende	Hornblende
per cent K	0.292	0.311
ppm rg ^{40}Ar	0.000 16	Below limit of detection
Apparent age	7.8 ± 1 Ma	Younger than 1.6 Ma

It is inferred that different parts of the diorite may have solidified at different times within the Pliocene–Pleistocene.

The Koloula Diorite may be subdivided into a series of imperfectly concentric zones, the boundaries of which may be gradational in a given area: the fresh tonalite or granodiorite core; a zone of hydrothermally altered quartz-diorite characterised by the introduction of secondary quartz and pyrite, widespread kaolinisation and some sericitisation; and hornblende melanodiorite and augite-gabbro which form an outer envelope or occur as 'rafts'.

Xenoliths of augite-gabbro and hornblende-diorite, also of altered basaltic andesite derived from the Suta Volcanics, are found in the granodiorite and quartz-diorite. Locally orbicular structures have developed around mafic nuclei.

The relationships suggest prolonged or polyphase passive emplacement by stoping, with periodic remobilisation of a very shallow acidic core. Clean contacts between the intrusive and the surrounding Suta Volcanics have not been observed, except in xenoliths; chilled contacts have not been found, although the adjacent area of Suta Volcanics has been hydrothermally altered. Along the line of the Toghonani Fault unaltered Suta Volcanics have been juxtaposed with the granodiorite core.

The core of the stock has been fractured by a regular joint system (Plate 13). Shearing along joints has formed zones of 'grush', which, in the hydrothermally altered zone, often pass into pyritic lodes.

A few narrow dykes of microdiorite or microgabbro intrude the Suta Volcanics in the lower Koloula and Kuma valleys; it is uncertain whether these should be associated with the Koloula dioritic emplacement or regarded as a late hypabyssal phase of the Suta vulcanicity.

Petrography

In hand specimen the granodiorite of the core is a medium- to coarse-grained leucocratic rock, in which quartz, plagioclase, hornblende and biotite are macroscopically conspicuous. In thin section the texture is hypidiomorphic-granular. Sixty per cent of the rock is made up of subhedral plagioclase, with individual crystals attaining a length of 5 mm. Albite and carlsbad-albite twinning and oscillatory-normal zoning are well developed. The composition of the plagioclase varies from An_{50} in the cores to An_{20} in the rims. The feldspars are not fissured as intensively as in the Suta Volcanics. Green hornblende ($2V \approx 60°$) comprises about 15 per cent of the rock, poikilitically enclosing small crystals of plagioclase, iron ore and apatite. Quartz, as anhedral crystals and aggregates, also makes up about 15 per cent of the rock; it poikilitically encloses both plagioclase and hornblende. Dark-brown biotite and iron ore, with traces of sphene, apatite and zircon comprise the remaining 10 per cent.

Towards the margin of the granodiorite core, the biotite has been converted to a golden-yellow chloritic mica; sparkling cleavage flakes of yellow chlorite are abundant in the river terrace gravels of the middle Koloula Valley.

In shear zones, quartz, chlorite and sericite form a higher proportion of the rock and cataclastic textures may be developed.

Hornblende-melanodiorites, which phase locally into augite-gabbro, are associated with the outer margin and roof of the diorite stock; the melanocratic rocks form only about 10 per cent of the diorite exposures mapped.

In hand-specimen the melanodiorite is usually finer-grained and more variable in appearance than the grano-diorite. A slight porphyritic tendency is often noted; in thin-section the texture appears inequigranular hypautomorphic. The coarse facies are often dark purplish-grey and green in colour, the green being due to clots of actinolite-chlorite intergrowth. These intergrowths have probably been derived from augite, as relics of clinopyroxene twinning are discernible. Skeletal iron ore is more abundant than in the granodiorite, but quartz makes up only about 5 per cent of the rock. There are rare patches of myrmekitic intergrowth of plagioclase and vermicular quartz.

Petrochemistry

Two analyses are available for rocks of the Koloula complex (Table 6). The analysis of a sample of fresh granodiorite from the lower Koloula Valley (65512) compares closely with Daly's average tonalite: this is the most acidic rock so far analysed from the Solomon Islands. The percentages of K_2O and total iron are somewhat low, although well within the ranges quoted for tonalite by Johannsen (1939, Vol. 2, p. 386).

According to Hatch and others (1949, p. 197) 'the centre point of the [granodiorite] family would average 66 to 67 per cent total silica, and of this about 22 per cent is quartz. There is a perfect gradation into tonalite, which differs only in the subordination of quartz.' The Koloula Valley contains over 27 per cent normative quartz, so that the term 'granodiorite' might be more appropriate for this particular analysis. Also Wells regards 66 per cent SiO_2 as the critical level to be used for separating diorite from granodiorite: on this basis also the rock must be regarded as a granodiorite. However, weight percentage of calc-alkali feldspar is

greater, and that of orthoclase less, than in Lindgren's average granodiorite (acid andesine 40, orthoclase 18). Normative proportions for 65512 give a plagioclase composition of An_{27}—a basic oligoclase.

A sample (65513) of a gabbroic facies of the dioritic stock from the Charichighi River is petrochemically an olivine-tholeiite. It is less alkaline than most of the basic Mbirao Volcanics, but shares more features in common with the pre-Miocene basic igneous rocks than either the Lower Miocene or Plio-Pleistocene volcanics. However, the K_2O/Na_2O ratio is 0.17—higher than in the basalts of the Mbirao Group. The copper content (220 ppm) is higher than that of the granodiorite, comparing with the arithmetic mean for the Mbirao Volcanics analyses (202 ppm).

The analysis of the Mbalisuna Gabbro (BSI 6459, Table 6) is closely comparable, except that the K_2O/Na_2O ratio is much lower—0.1.

Examination of thin sections has shown that there is considerable petrological variation within the Koloula stock. According to Joplin (1968) the heterogeneous composition of many diorite bodies lends support to the thesis that they originate in depth as hybrid magmas formed by the reaction between country rocks, possibly of basaltic composition, and granitic fluids. The presence within the granodiorite of mafic xenoliths, schlieren and raft-like bodies of melanodiorite or alkali-gabbroic material strongly suggests that this kind of hybridisation has occurred. The basic material may represent in part ingested basalts of the Suta Volcanics suite; the coarser gabbro may be of remobilised derivative of the pre-Miocene floor, comparable to the Guadalcanal Gabbro of the Sutakama–Sutakiki area 15 km to the north.

The granodiorite of Koloula might be considered the most acid representative of a calc-alkaline trend which gave rise to tholeiitic basalts (and basaltic andesites) in the Lower Miocene (the Suta Volcanics) and hornblendic andesites in the Pliocene (Gallego Lavas). Clearly the petrology of the Koloula stock could prove to be very complex in detail; more rock analyses and further petrographic work are re-

Plate 13 Jointing in the Koloula Diorite, Kolochoro tributary of the Toghonani River, south-central Guadalcanal.
The granodiorite has been sheared in parallel zones due to movement along a set of regular tension-joints

quired before generalisations can be made about the major trends or mode of emplacement.

Geothermal springs

Figure 7 shows 24 hot-water springs in relation to major faults and the Plio-Pleistocene igneous rocks of Guadalcanal. Thirteen of these are little more than cold sulphurous or saline springs which have been used by local people as sources of 'tasi' (='salt' or 'sweetness'). These phenomena are restricted to a belt up to 15 km wide, which extends from the Savulei area south-east to Veuru Moli; they are associated in part with areas of Plio-Pleistocene vulcanicity and complex fracture zones trending north-east.

THE HONIARA BEDS AND ASSOCIATED SEDIMENTS

Coleman (1957) gave the name Honiara Beds to 'the calcareous sediments of varied lithology which rise from the sea as a series of three or four terraces . . . and which occupy a belt of country from the Umasani (Segilau) River to the Lungga River, and beyond'. Hackman (1968c) introduced the name Longgu Beds for 'up to 200 feet of variable calcarenite, ortho- and para-conglomerate and brecciated coral reef' outcropping in the hinterland of Kaoka Bay. The

Mbota Moli Beds (Hackman, 1968c) outcrop discontinuously in south-east Guadalcanal between Cape Henslow and Kopiu Bay. These three rock units are here treated together, as they show similarities in lithology, fossil content and stratigraphic position.

The Honiara Beds form a grassy escarpment broken by all the major rivers between the Umasani and the Mbokokimbo: behind Honiara itself there is a fine series of terraces. The Longgu Beds form a forested escarpment which rises 400 m above the Simiu Valley and slopes gently eastwards to Kaoka Bay. The Mbota Moli Beds appear as a smoothly textured veneer to a bevelled platform of Mbirao Volcanics.

The Vara Creek tributary of the Mataniko provides a fair type section for the Honiara Beds. Across the eastern part of the outcrop, the Ghauregha tributary of the Mbalisuna provides a continuous section.

The Longgu Beds (Ql) are well exposed in the Mbo'o River, whilst the Mbota Moli Beds (Qm) are exposed discontinuously in the Mbalo River.

The Honiara Beds have a rich fauna: molluscan remains are more abundant than in the older sediments. Noteworthy is the giant clam, *Tridacna*. Partly recrystallised coral-algal material is a common constituent of the limestones. The benthonic foraminifera *Calcarina spengleri* and *Baculogypsina* are found in the Honiara Beds and the Mbota Moli Beds. Large numbers of *Operculina* individuals have

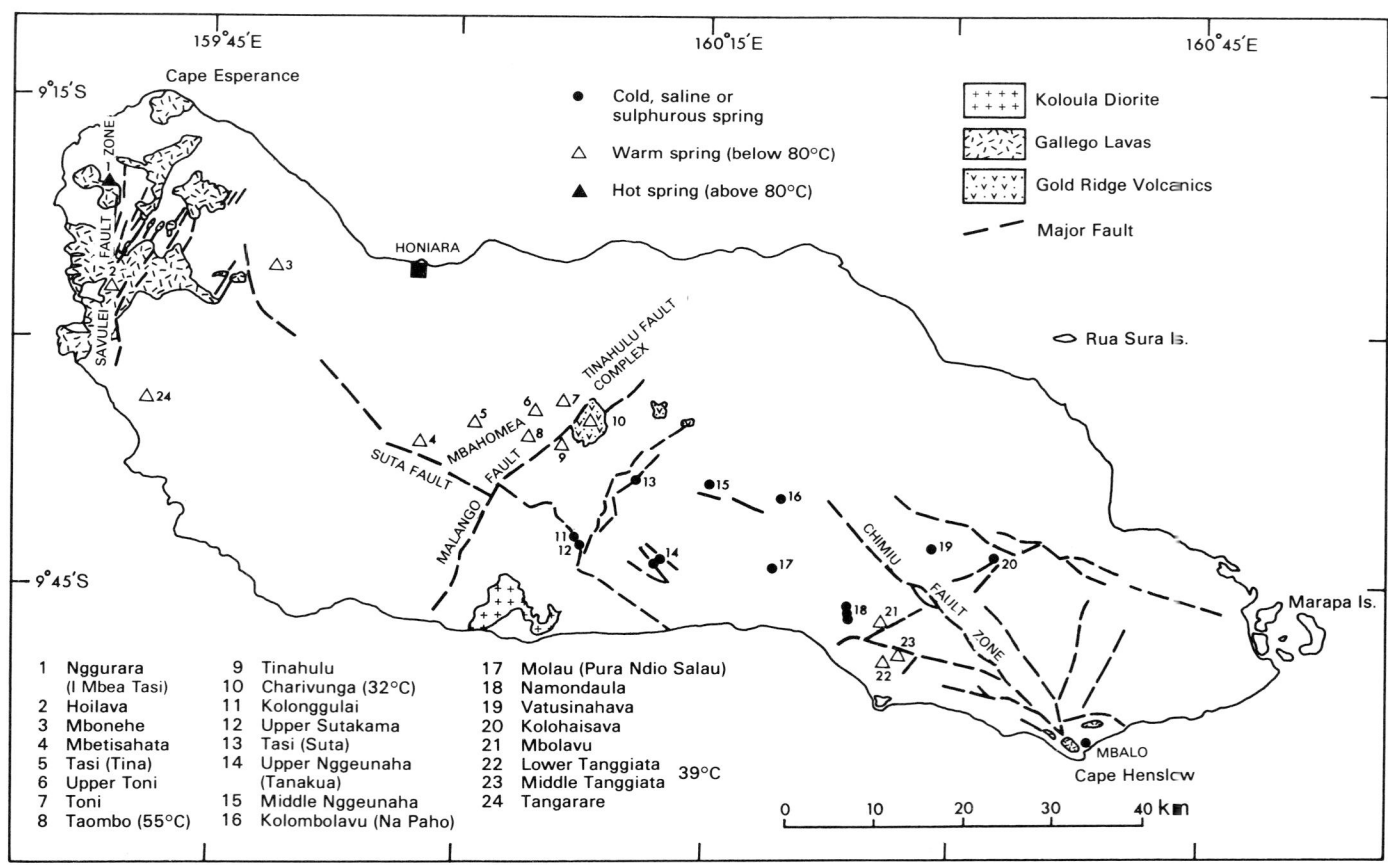

Figure 7 Map showing the location of geothermal springs in Guadalcanal

been found in the basal part of the Longgu Beds. Planktonic foraminifera and carbonised wood fragments are common in the sandy facies.

The fauna and well-preserved terraces indicate a Recent or Pleistocene date for the Honiara and Mbota Moli Beds. Foraminifera from the Longgu Beds indicate a Pliocene–Quaternary age (P. J. Coleman, written communication, 1969).

The Honiara Beds rest disconformably on the Lungga Beds and Mbonehe Limestone to the west of Honiara, and overlie various members of the Toni Formation to the east. They interfinger with the alluvials of the Guadalcanal Plains and the upper part of the Lungga Beds. The bedding dips at 5° to 20° to the north-north-east. The Longgu Beds and Mbota Moli Beds rest unconformably on the Mbirao Group: there are local variations in dip due to faulting and differential compaction over an irregular basement.

The maximum thickness of the Honiara Beds, between the Lungga and Mberande Rivers, is about 800 m, but they thin rapidly to about 60 m west of Honiara. The thickness of the Longgu Beds does not exceed 300 m; the Mbota Moli Beds are about 200 m thick.

In the Honiara Beds arenaceous and rudaceous epiclastics bulk large, as in the underlying Toni Formation, but calcareous sediments form a higher proportion of the sequence, especially west of the Lungga River.

Conglomerates predominate towards the eastern end of the Honiara Beds: they comprise the lower two-thirds of the section in the Ghauregha tributary of the Mbalisuna. Orthoconglomerate passes up-section into grey-weathering rusty yellow arenites with occasional pebbly bands. Seventy per cent of the phenoclasts in these conglomerates are of hornblende-andesite of Gallego type; fragments of feldsparphyric basaltic andesite and coralgal limestone are also common. Conglomerates also occur in the Longgu and Mbota Moli Beds; in the former case, the phenoclasts are derived from the Mbirao Group, in the latter they include fresh andesite derived from the Cape Henslow area.

The arenites tend to be poorly sorted, with a high carbonate content: Coleman (1957) used the term 'calcareous tuff'. Angular crystal fragments of basaltic hornblende may comprise up to 65 per cent by volume of the whole sediment. Similar rather incoherent arenites make up about 60 per cent of the Longgu and Mbota Moli sequences.

Coralgal limestones, relatively uncontaminated by terrigenous or pyroclastic material, occur as poorly-bedded discontinuous lenses within the arenites, particularly in the terraced outcrop behind Honiara town. The uppermost 60 m in this area forms a distinct complex of uplifted coralgal reef, derived rubble and calcirudite, the Honiara Reef Limestones. A sharply differentiated biohermal structure, the Gharangi Knoll, also occurs about 3 km south of Tutumu village: this is an elliptical knoll about 3 km long and trending north-east. The airborne scintillometer survey (ABEM, 1967) recorded increased radioactivity over this knoll, coinciding with a magnetic anomaly in depth; possibly a buried extension of the Mbalisuna Gabbro formed a suitable substratum for reef growth. The radiometric anomaly may be attributed to uranium in the red clay which characteristically develops in pockets on uplifted Pleistocene–Recent limestones throughout the Solomons.

The lithological association is essentially a complex of biogenic reef and estuarine or shallow neritic sediments; they may be regarded as recently elevated offshore facies.

RAISED PLEISTOCENE TERRACES

In certain parts of the foothills zone of central Guadalcanal, the finely dissected interfluves have been bevelled by erosion surfaces which post-date the Pliocene sediments. On the aerial photographs they show as distinct patches up to 4 km² in extent, notably in the areas of Kolochulu, Mt Ndatovitu and the Haviha Ridge. Figure 6 (Profile AA′) shows how the surface at Kolochulu truncates the topography of the foothills. Even the bevelled summit of Mt

Plate 14 Gravity folding in Recent laminated lacustrine silts, Na Uli tributary of the Valasi River, east-central Guadalcanal

Tatuve, which exceeds 1400 m above sea-level, could be a remnant of one of these surfaces.

The surfaces slope at very low angles to the north: no littoral deposits have been found in association with these areas, which are forested and deeply lateritised.

Conceivably these terraces were bevelled by marine erosion in the late Pliocene or early Pleistocene. The terrace sequence which is so well displayed behind Honiara comprises at least six distinct levels of marine erosion.

Mitchell (1969) on the basis of a detailed survey of terraces in North Malekula, New Hebrides, explained the sequence in terms of the interaction of progressive uplift and Pleistocene eustatic fluctuations in sea-level. The Guadalcanal terraces might be explained satisfactorily on similar lines, indicating that the island has been uplifted *at least* 800 m relative to sea-level since the end of the Pliocene.

RECENT SEDIMENTS

Alluvials Coleman (1960b) described the outwash material which floors the deltaic alluvial plains of northern Guadalcanal under the term 'Ngalimbiu Alluvials'. The surface of the plains dips at a very low angle northwards; river bluffs provide occasional sections in immature polymictic gravels, with lenses and shoestrings of sand and silt. The larger phenoclasts are of boulder size, usually well rounded, and predominantly composed of material derived from the Pliocene sequence of the foothill zones.

The small alluvial area of Kavahambe (Kovagombi) is of particular interest in that it contains gravel beds about 13 m thick with moderately high gold content, presumably derived from the Gold Ridge area. Stanton (1965), describing this small pocket of coarse estuarine sediment, remarks on the difficulty of determining 'whether certain river beds are composed of Tertiary material or simply recent semi-consolidated river sand and detritus'; this close similarity between the Toni Formation and the alluvials suggests that their depositional environments were, in part, comparable.

The south coastal strip is composed of sandy gravels and outwash from the Mbirao Mountains: west of Avuavu it is little more than a storm beach.

Lacustrine deposits In the mountains there are occasional deposits of Recent blue silt forming level areas where the main river valleys widen above gorge-like constrictions. Noteworthy is the development of silt in the Valasi River around the village site of the same name; in the Na Uli tributary a section in finely laminated silt reveals a remarkable sequence of overturned folds, indicating post-depositional slumping towards the centre of a small lake-basin (Plate 14).

Coral reef Reefs fringe the coastline, particularly in the area of Marau Sound; there is very little active growth at the present time. Rua Sura and Nudha Islands are cays of skeletal coral sand; the reefs and sand cays of the Marau area have been described by Stoddart (1969a and b).

THREE

The structural geology of Guadalcanal

The tectonic domains map of Guadalcanal (Figure 8) synthesises all the structural information available, derived from the following sources: field descriptions of mesoscopic structures (Turner and Weiss, 1963, p. 15); photogeology; and geomorphological features.

Field data

In central and eastern Guadalcanal the attitudes of the following structural features were recorded in the field:
1 Bedding planes and original foliations in igneous bodies (Figures 9, 13).
2 S-surfaces, i.e. planes of schistosity, cleavage or shear (Figure 10).
3 Minor folds (Figure 14A).
4 Fault planes, with evidence for sense of differential movement (Figures 17, 18).
5 Joint planes (Figure 19).
6 Mineralised veins (Figure 21).
7 Discordant minor intrusives (Figure 20).
The entire area has been subdivided with a view to relating this information to the framework of the major mappable units. Some of these units have been further subdivided where the volume of data available so warrants, for example in the case of s-surfaces in areas 5, 6, and 7.

The following scheme has been used:
Area 1   Weather Coast Block (west)
Area 2   Weather Coast Block (east)
Area 3   Valasi Block
Area 4   Suta–Malango Area
Area 5   West Mbirao
Area 6   Central Mbirao
Area 7   East Mbirao
Area 8   Koloula–Haiacha and the Itina Valley
Area 9   North-central sediments
Area 10   North-eastern sediments

Photogeological data

The photogeological work has contributed the following elements to the structural picture: coarse lineaments, representing in most cases major faults or fault zones; fine lineaments, the expression of sets of penetrative s-surfaces; and areas of uniform dip in the sedimentary succession, escarpments, etc.

Bedding planes in the Mbirao Volcanics

The poles to 118 readings of pre-Miocene bedding inclination have been plotted on the lower hemisphere of a stereographic projection, the Schmidt Equal Area net (Figure 9). This is a small number of observations considering the size of the area: it reflects the difficulty of finding exposures which can be interpreted with a fair degree of certainty.

Within the Mbirao Volcanics the attitude of the stratification may be estimated where pillow structure is well developed and the jointing allows a reasonably good three-dimensional picture to be obtained. Tuffaceous bands and flow structures which might give some indications of bedding are only rarely encountered.

Suitable exposures can be studied on the western side of Kopiu Bay on the south-east coast, where the lava pile is inclined at about 30° towards true north. Elsewhere within the Weather Coast Block there is a tendency for readings to cluster between 10° and 50° towards the north-north-east; Figure 9 shows a distinct maximum for the whole area at 25°–20°. There is a wide scatter of points, which is to be expected in view of the difficulties of estimating inclinations from pillowed outcrops, and taking into consideration the localised affects of faulting.

Rather steep northerly dips have been recorded in the lower part of the Sambahalava River, possibly reflecting a northerly tilt associated with down-faulting to the south, along the line of the Veuru Moli Fault (Figure 6, DD′).

Along the northern margin of the Weather Coast Block there is some consistency in the direction of dip: the angle varies between 8° and 35° towards the north-north-east (in the headwaters of the Kolohaisava River), while to the east

in the heads of the Kolovaghamela and Mandonu rivers the dips are steeper, varying from about 25° to 85° to the north or north-north-east. To the west, at the mouth of the Talise River, the stratification is nearly vertical; again this is probably due to Recent down-faulting to the south, parallel to the coast.

In the Valasi Block pillow formation is not developed to the same extent as in the Weather Coast Block, and within the Mbirao Metabasics the difficulties of observation and interpretation are accentuated by the greater intensity of alteration, shearing and faulting. However, the Tetekanji Limestones do provide stratification markers where they are interbedded with the basalts: in the headwaters of the Kolohaisava Rivers there is a tendency for these small limestone bodies to strike north-west and dip at 60°–70° to the north-east or south-west. Two isolated readings from the metabasics west of the Sutakama Valley conform with the north-north-easterly maximum.

The overall pattern on the composite contoured density diagram merits comparison with patterns for s-surfaces in adjacent areas of the Metabasics belt, e.g. sub-areas 6A and 7A on Figure 10. The distribution on Figure 9 defines a great circle indicating regional folding about axes plunging approximately 15° to the north-west.

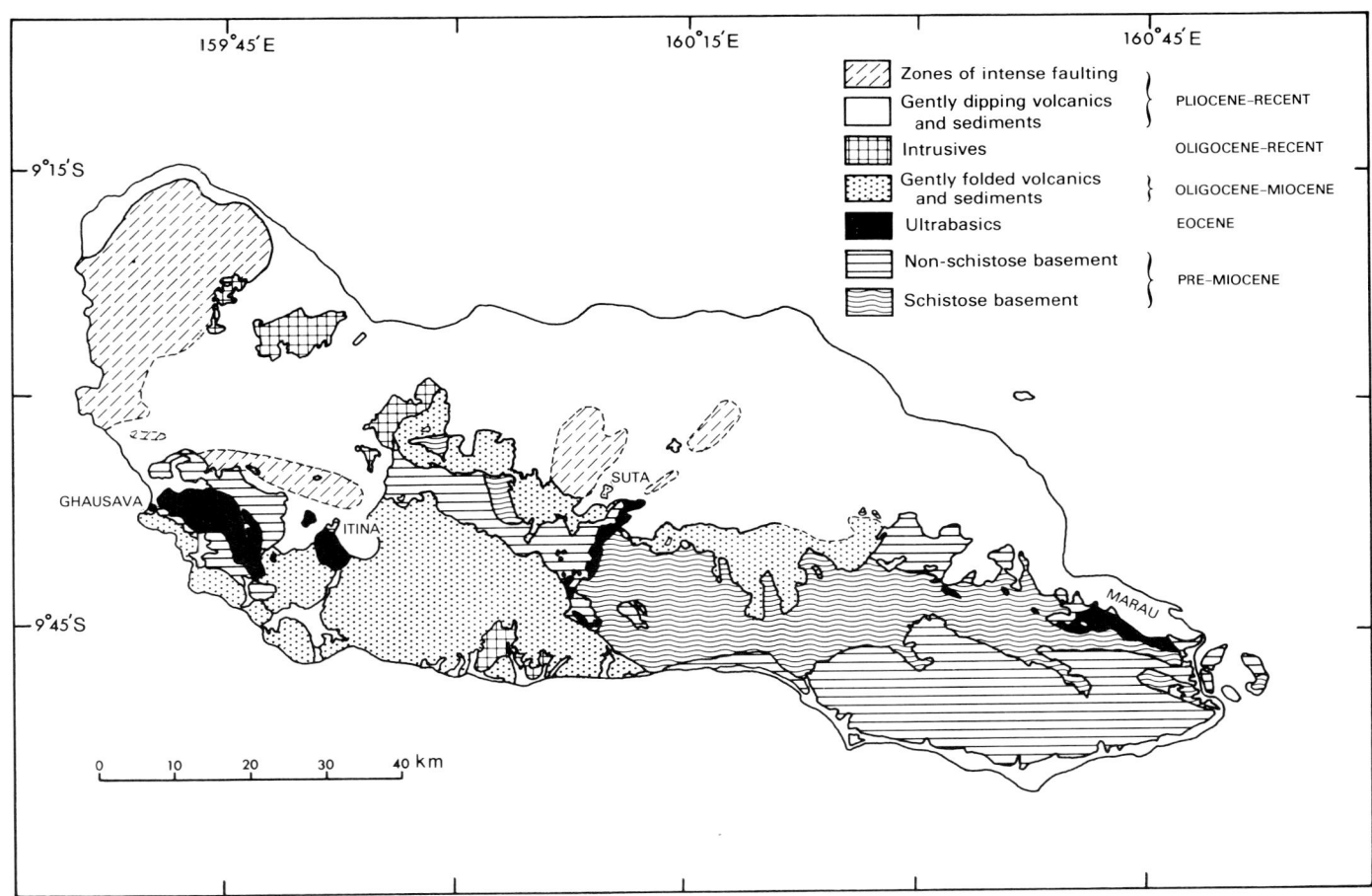

Figure 8   Tectonic domains of Guadalcanal

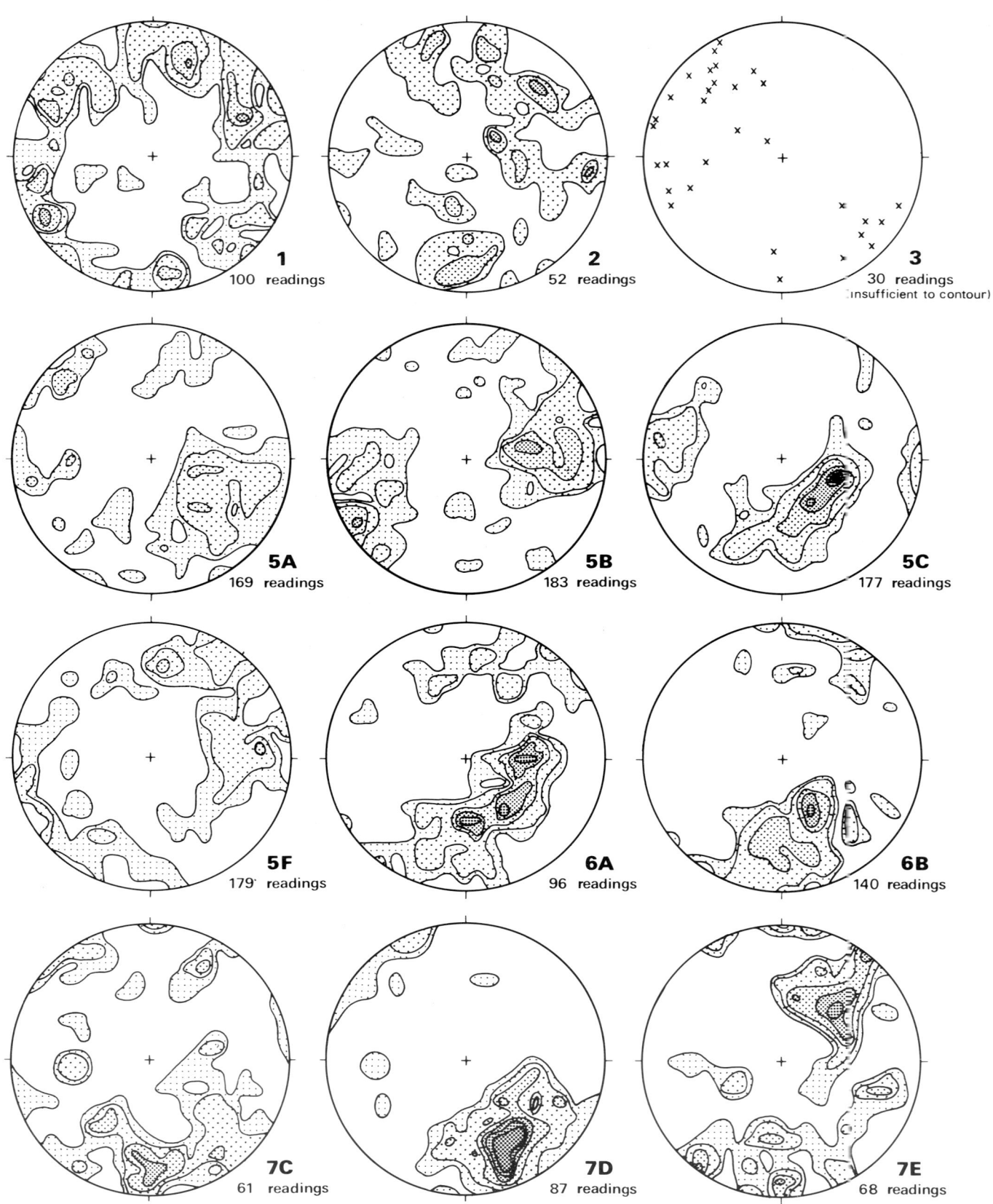

Figure 10 Contoured equal-area stereographic projections of poles to metamorphic s-surfaces in the pre-Miocene of Guadalcanal

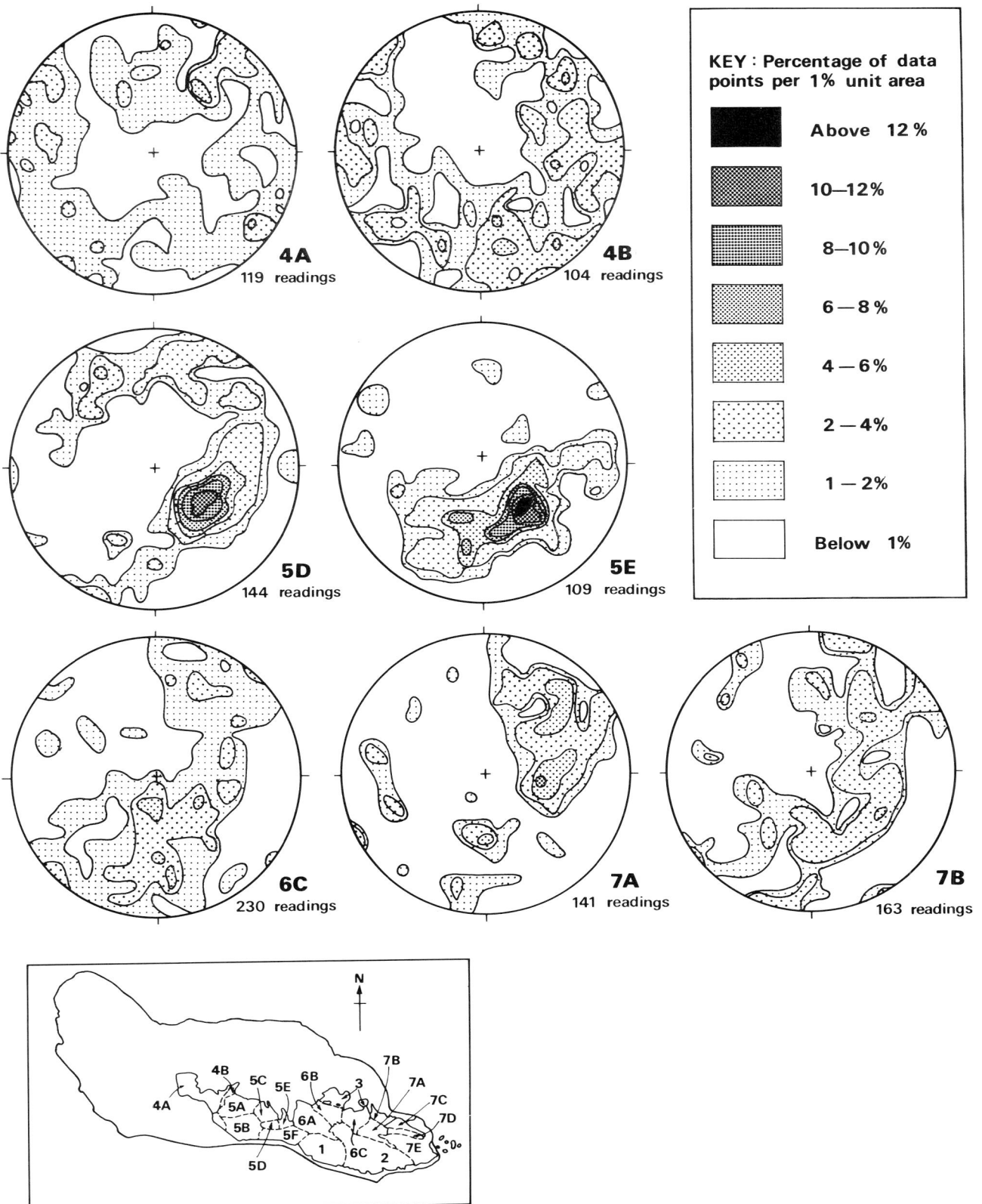

KEY : Percentage of data points per 1% unit area

Above 12 %

10—12%

8—10%

6—8%

4—6%

2—4%

1—2%

Below 1%

4A
119 readings

4B
104 readings

5D
144 readings

5E
109 readings

6C
230 readings

7A
141 readings

7B
163 readings

Metamorphic s-surfaces

To avoid genetic implications, the term s-surface has been used in preference to schistosity, foliation or cleavage. (Turner and Weiss, 1963, p. 28).

Within the Mbirao Group the attitudes of 2322 s-surfaces have been recorded, 2021 of these from within the Mbirao Metabasics belt. Poles to these surfaces were plotted on the Schmidt Equal Area Projection, and the contoured diagrams are presented in Figure 10.

No previous attempt has been made to map s-surfaces systematically within the Mbirao Group, although Thompson (1960) described a 'cleavage' in the Guadalcanal Schists on the southern flank of the Marau Ultrabasics. He produced a stereographic plot of the cleavages which showed a distinct concentration of surfaces dipping to the north-west at fairly low angles, i.e. in general agreement with the pattern for sub-area 7D of Figure 10.

The boundaries of the areas and sub-areas were delineated as far as possible to coincide with the broader structural divisions, for example the Valasi Block (area 3), and also to

coincide with areas where there appeared to be some degree of homogeneity of pattern. This was assessed subjectively after the data had been plotted on a scale of 1:50 000. For example, a boundary was drawn between sub-areas 5E and 5F on the basis of the apparent predominance of a north-easterly tread to the s-surfaces in the former and a north-north-westerly trend in the latter; in this case the zone of transition between the two sub-areas might be considered to pass across the upper valley of the Alivaghato River.

The s-surfaces mapped by the writer can be considered in terms of three idealised categories:

1 A penetrative planar fabric or well developed 'schistosity' defined by parallelism of tabular crystals of chlorite, fine-grained silky mica, or fibrous aggregates of actinolite.

2 A less penetrative type of crenulation foliation (Whitten, 1966, p. 232) or crenulation cleavage (Knill, 1960) which has been superposed on a more penetrative fabric; slip is visible on individual planes, at least in thin section if not macroscopically.

3 A cataclastic cleavage (Knopf, 1931) in which there appears to be a planar fabric, usually penetrative over only narrow zones not exceeding 20 m in width, conditioned by a well developed cataclastic fabric.

In practice there appear to be all gradations between each of these categories, depending essentially on the amount of recrystallisation that has accompanied the shearing.

In general penetrative planar fabrics are developed towards the northern margins of the Mbirao Metabasics belt.

In Figure 10 the Mbirao Metabasics belt, i.e. areas 5, 6 and 7, has been divided into 13 sub-areas. Data for the Suta Ultrabasics and the Marau Ultrabasics have also been presented separately (sub-areas 4B and 7C respectively).

The stereogram for area 3 (the Valasi Block) has not been contoured for density distribution on account of the limited number of readings available.

Areas within the Metabasics belt which have one well-developed schistosity with consistent direction of strike, show up on the aerial photographs as finely scored by a set of very closely spaced parallel lineaments. A good example is to be seen in the upper valley of the Mandonu River, in eastern Guadalcanal, where the schistosity trends north-west and dips consistently at a fairly high angle to the south-west (sub-area 7E). Sub-area 5D, in the upper Talise Valley shows finely textured lineaments, trending north-east, parallel to a schistosity which is dipping at a low angle to the north-west. In sub-area 5A, the same lineaments are particularly prominent on the south-east flank of the Nggeunaha Valley: in plan the re-entrants point down the north-westerly flowing tributary valleys, i.e. parallel to the direction of dip (Plate 2). In adjacent areas, however, where two or more 'schistosities' may have been imprinted, the pattern of lineaments on the aerial photographs is not so clearly defined.

The sequence of s-surfaces

The headwaters of the Kolombolavu River, within sub-area 5E, probably afford the best sections for examining the relationships between different generations of s-surfaces.

Figure 9 Contoured equal-area stereographic projection of poles to bedding planes in the Mbirao Group

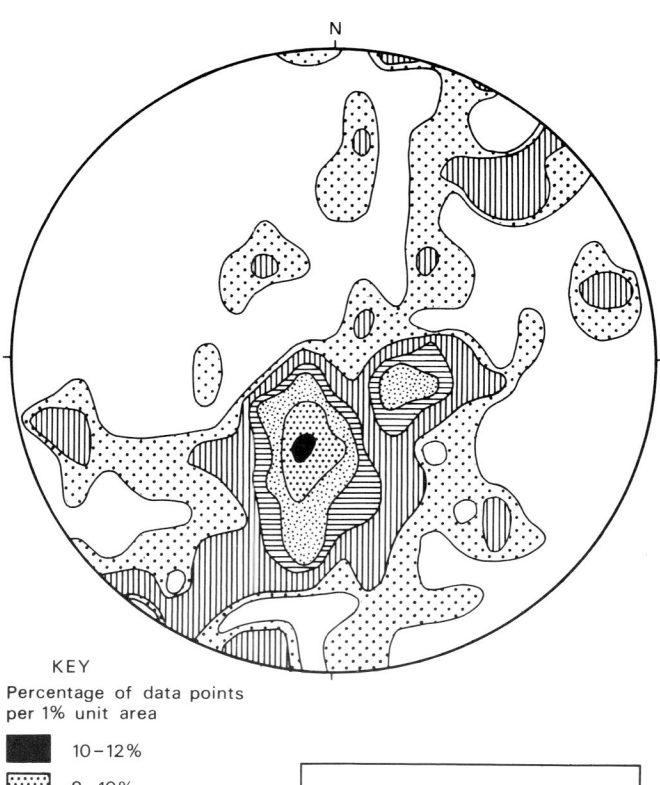

KEY

Percentage of data points per 1% unit area

■	10–12%
▨	8–10%
▦	6–8%
▤	4–6%
▥	2–4%
⚬	1–2%
□	Below 1%

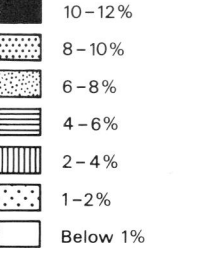

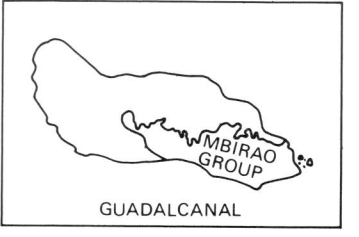

GUADALCANAL

Here the following sequence can be clearly demonstrated:

1 S_1: A penetrative planar fabric in chlorite-schist, dipping at approximately 35° towards the north-west.

2 S_2: A planar structure defined by a crenulation or strain-slip type of cleavage, dipping at about 50° to the north or north-north-east.

3 S_3 and S_4: Poorly developed cataclastic cleavages, dipping at about 50° to the east-north-east. The trace of these surfaces on S_1 gives a faint lineation which plunges consistently at about 35° towards the north-west. This lineation is parallel to the axes of gentle folds of low amplitude which have affected the S_1 surfaces.

Using this sequence as a working model, it is a fair assumption that the orientation of the later s-surfaces should be reasonably consistent over the entire structural domain in which constant stress-strain relationships held sway, whereas the earlier s-surfaces would be expected to show some disturbance in orientation pattern due to their accommodation to later stresses. In fact in sub-area 5E the density maxima corresponding with S_1 are more widely dispersed than those for S_2, and could be considered to form part of a girdle consonant with rotation about an axis parallel to the strike of S_3.

The correlation of s-surfaces from sub-area to adjacent sub-area is not a straightforward matter, as the full sequence is not everywhere equally well developed. Although the Mbirao Group as a whole is petrologically homogeneous, one would expect the pattern of strain at higher levels in the lava pile to differ from that of the deeper levels which were undergoing regional metamorphism.

In Figure 11 the maxima have been extracted from the stereographic plots and presented in the form of trend diagrams, to give some impression of the possible correlations between the sub-areas. There is a good measure of parallelism between trends along the northern margin of the metabasic belt, in the sense that s-surfaces dipping at fairly low angles towards the north-west dominate the structural picture. East–west trends are important in the central and eastern areas of the metabasic belt, but there is an abrupt change in pattern on the south side, north-westerly trends predominating in sub-areas 5B, 5F and 7E.

Table 8 shows where sequences have been established unequivocally from examination of field relationships. Although the relationships between three sets of s-surfaces have been clearly demonstrated in only two sub-areas, i.e. 5E and 7E, in each case where a cross-cutting relationship has been recognised, east–west-trending s-surfaces post-date those dipping at shallow angles to the north-west, and north-westerly-trending s-surfaces post-date the other two; the latter are occasionally found parallel to the axial planes of minor flexures which affect the earlier schistosities.

S_1 surfaces Throughout the northern half of the Mbirao Metabasics belt, the most conspicuous s-surfaces tend to dip at fairly low angles, 25°–50° to the north-west.

The density plot for S_1 forms a very distinct girdle in sub-areas 5E, 6A and 7A; the maxima in the south-east quadrant show a similar tendency in sub-areas 4B, 5A, 5C, 5D, 6C, 7B, 7D and 7E. In each case the axis of rotation plunges at a shallow angle to the north-west. Over the whole belt there is a variation of about 70° in the strike from north-north-east to east-north-east. There is a tendency for the north-westerly angle of dip to increase the more closely the strike of the s-surfaces approaches an east–west trend, reaching maxima of over 70° in the extreme east.

Parallel trends are poorly developed in sub-area 4A and along the southern coastal fringe of the Weather Coast Block; passing south into sub-areas 5B and 5F the penetrative 'schistosity' gives way to an impersistent cataclastic or shear cleavage. In sub-areas 1, 4A and 5B a pair of conjugate shears has developed, and in sub-areas 2 and 5F the surfaces dip at steeper angles to the south-east, antithetic to parallel trends to the north.

S_2 surfaces The density maxima for S_2 are not dispersed to the same extent as those for S_1.

The prevailing trend is east–west or east-south-east, the schistosity being most strongly developed in the central and eastern part of the Metabasics belt. Predominantly, S_2 planes are inclined at 40°–50° to the north, but there is a wide variation in the amount of dip.

On the stereographic projections the maxima for S_2 are in general more strongly concentrated than for S_1, there being little tendency to develop a girdle.

S_3 and S_4 surfaces The two trends interfere in sub-areas 5B and 7C, where the north–south schistosity clearly post-dates an earlier north-westerly trend. The density maxima are in general more clearly defined than in the diagrams for S_1 and S_2. However the distinction between S_3 and S_4 is by no means clear in all sub-areas; S_3 and S_4 might be considered as manifestations of a third phase of deformation, but their separation and correlation between sub-areas should be regarded as tentative only. In the northern part of the Mbirao Metabasics belt S_3 and S_4 are impersistent and but poorly developed, although they are consistently axial to minor folds in S_1, especially in sub-areas 5E and 6B. A more penetrant schistosity, with recrystallisation and annealing of shear surfaces, appears in sub-areas 5B and 5F, also in the upper Mandonu Valley in sub-area 7E in the extreme east, where the s-surfaces dip consistently at about 50° to the south-west.

Table 8 List of localities where sequences of metamorphic s-surfaces can be observed

Sub-area	Locality	S-surface			
		S_1	S_2	S_3	S_4
4A	Mbeambea R.	×	−	×	−
5A	Nggeunaha R.	×	×	−	−
5B	Alitauva R.	−	−	×	×
5D	Talise R.	×	×	−	−
5E	Hailava R.	×	×	−	×
6A	Upper Rere R.	×	−	×	−
6B	Upper Rere R.	×	×	×	−
6C	Upper Na Ndoli R.	−	×	×	−
7A	Middle Lughumboko R.	×	×	×	−
7E	Kolovaghamela R.	−	×	×	×

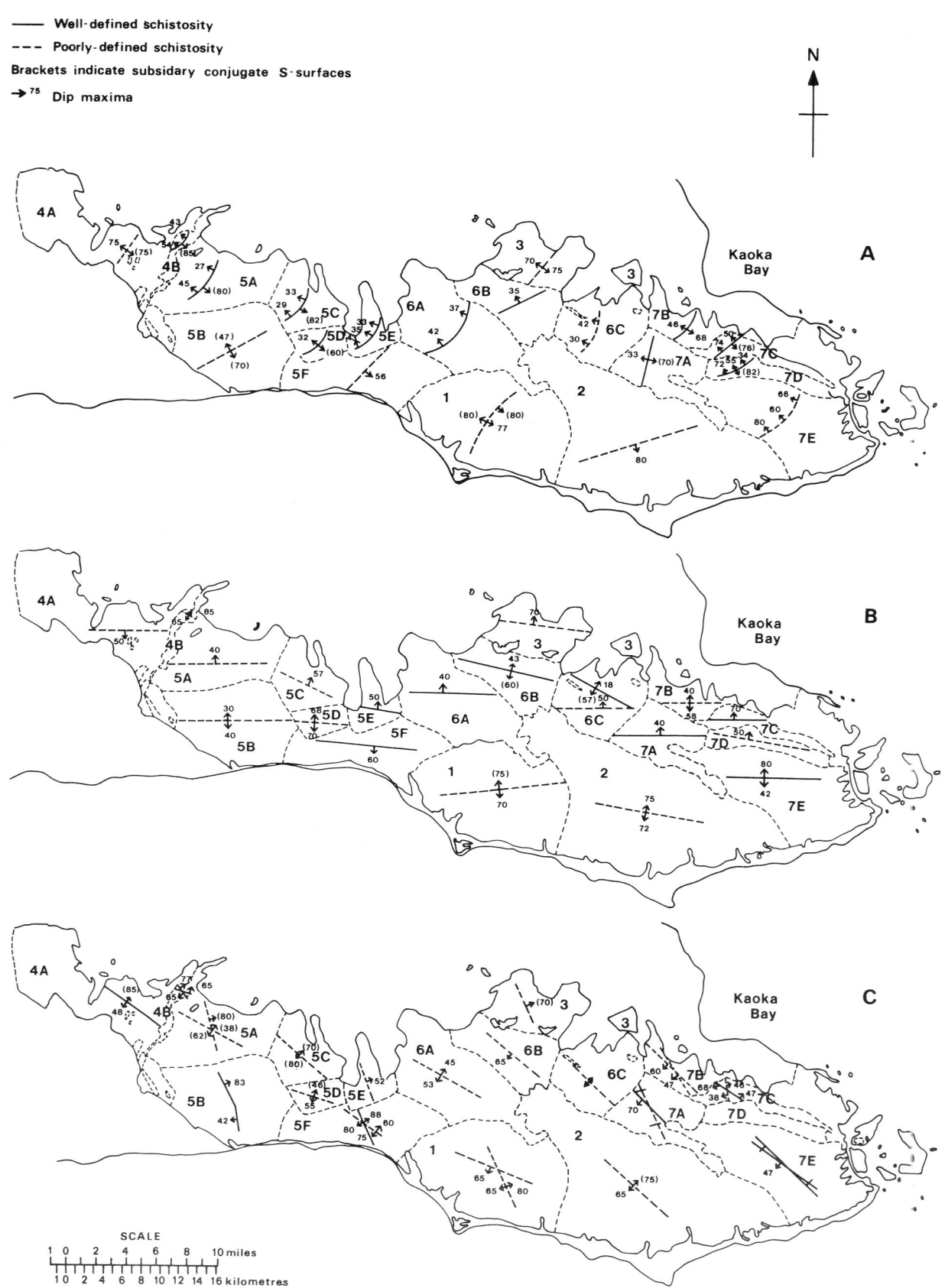

Figure 11 Maps showing trends of metamorphic s-surfaces in south-east Guadalcanal

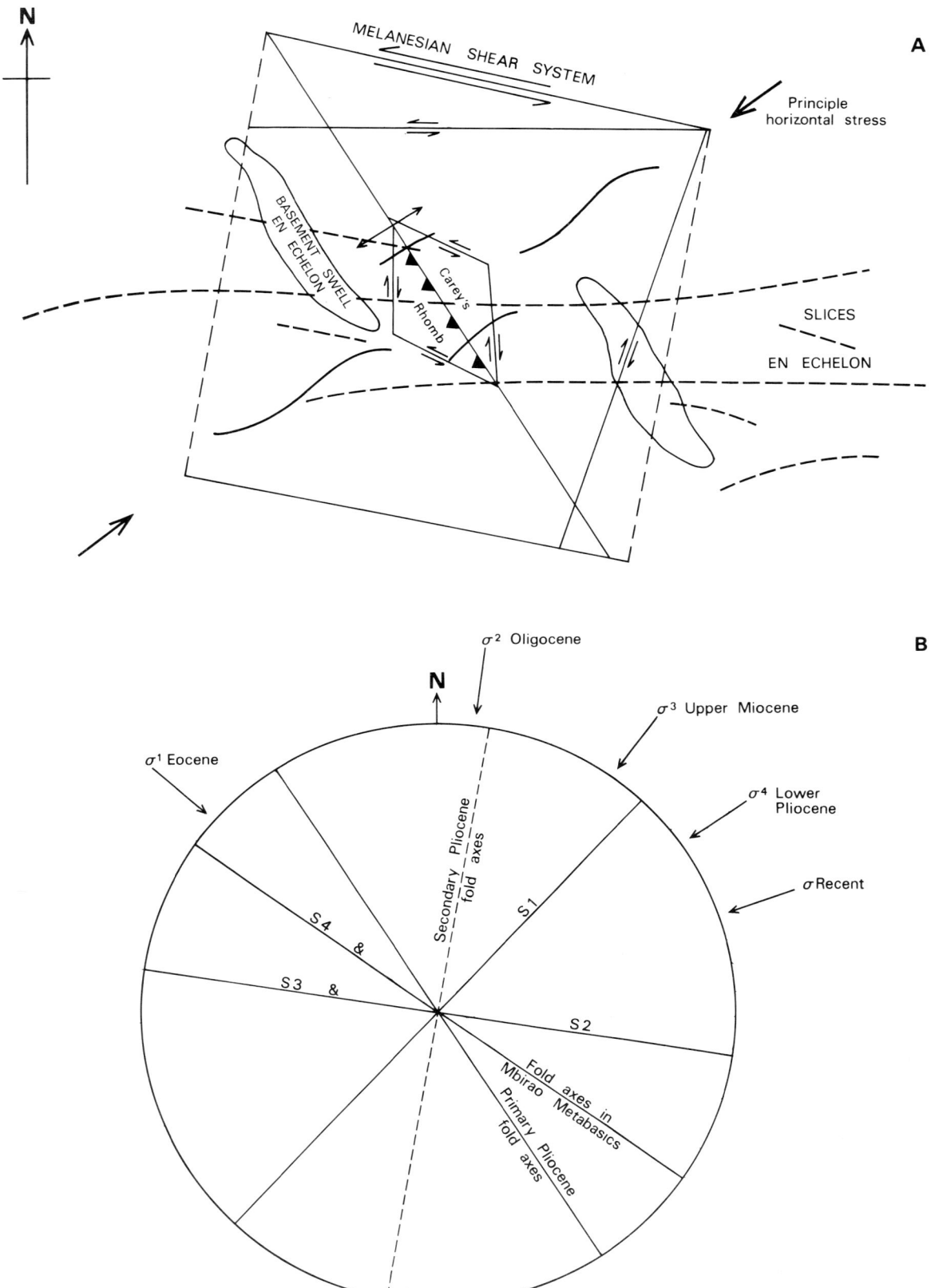

Figure 12 Changes in the direction of maximum stress as deduced from patterns of s-surfaces in Guadalcanal

The significance of s-surfaces within the Mbirao Group

The existence of at least three, possibly four separate systems of s-surfaces with varying orientations indicates that the Mbirao Group has undergone a complex history of polyphase deformation.

However, the stress-strain relationships in a deformed pile of basic volcanics are not directly comparable with those in an area of deformed metasediments, for example where slaty cleavage in the pelitic members is oriented parallel to the axial planes of the folds and parallel to the AB plane of the strain ellipsoid (Wilson, 1946). The formation of schistosities in low-grade metabasic volcanics has not been dealt with in any detail in the literature. The original mechanical anisotropy in a pile of pillowed or massive lavas would be negligible compared with that of a sequence of interbedded psammites and pelites. During folding the role of flexural slip would be correspondingly minimal. It might be expected that in the upper part of the Mbirao Volcanics pile, the stress theory of rupture would be more applicable than in the metamorphosed infrastructure where the greater plasticity would be more in accord with the style of the strain theory (Wilson, 1946). Considering the petrologic homogeneity of the Mbirao Volcanics, it might be assumed that the style of deformation resulting from a particular phase of regional stress would depend only on the degree of metamorphism and proximity to active 'master structures'.

The 'cataclastic' nature of the s-surfaces, in particular of S_3 and S_4, and the development of conjugate pairs, for example in areas 1 and 2 of the Weather Coast Block, suggest that the later non-penetrative s-surfaces are more likely to correspond with the theoretical loci of stress shear planes, i.e. they are comparable to the shear cleavages of Wilson (1946). In such cases, according to stress theory, the derived axis of maximum stress should bisect the obtuse angle between the conjugate shears.

There is occasional evidence for transcurrent movement, commonly of the oblique-slip type, acting within the plane of s-surfaces in the Mbirao Group. However, in this respect the sense of movement is not consistent; furthermore S_3 is demonstrably axial to minor folds which affect S_1, and S_4 is parallel to the derived fold axes in the Miocene sediments (Figures 11c, 14b). Assuming that the schistosities were produced by non-rotational force acting on fairly homogeneous material, and that the directions of the principal axes of strain coincide with those of the principal stresses, then the s-surfaces are either normal to the maximum regional stress responsible, or, in the case of conjugate shears, the trace of their intersection is normal thereto.

In terms of the stress-strain relationships so conceived, the deduced directions of maximum regional horizontal stress for the sequence of s-surfaces within the Mbirao Group are, in relation to the present-day orientation of Guadalcanal, as follows:

S_1 North-west to south-east
S_2 North-north-east to south-south-west
S_3 & S_4 Approximately north-east to south-west

Accordingly the derived direction of maximum horizontal stress migrated progressively clockwise through an angle of approximately 90° during the tectonic history of the Mbirao Group (Figures 12b and 33).

Bedding planes in the Neogene sequence

Planes of stratification are well-developed in the Lower Miocene calcareous sediments, but are poorly defined in the Suta Volcanics; also, in the Middle Miocene to Pleistocene sequence the siltstones and fine-grained sandstones show plentiful bedding surfaces; however, the attitude of the stratification is more difficult to determine satisfactorily in the conglomeratic and pyroclastic facies.

The attitudes of a total of 1801 bedding planes have been measured from within the post-Oligocene sedimentary and volcanic sequence. The poles to these bedding planes have been plotted on the lower hemisphere of a Schmidt Equal Area projection, for each of 10 sub-areas (Figure 13).

Figure 13 Contoured equal-area stereographic projections of poles to bedding planes in the post-Oligocene sediments of central and eastern Guadalcanal

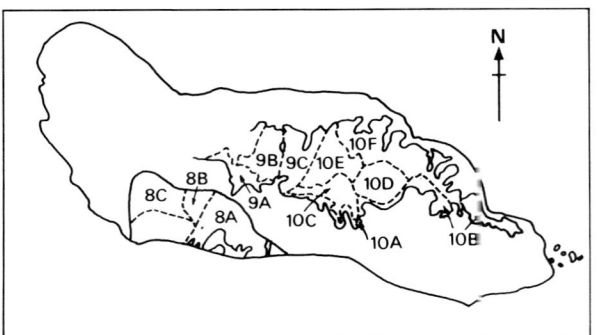

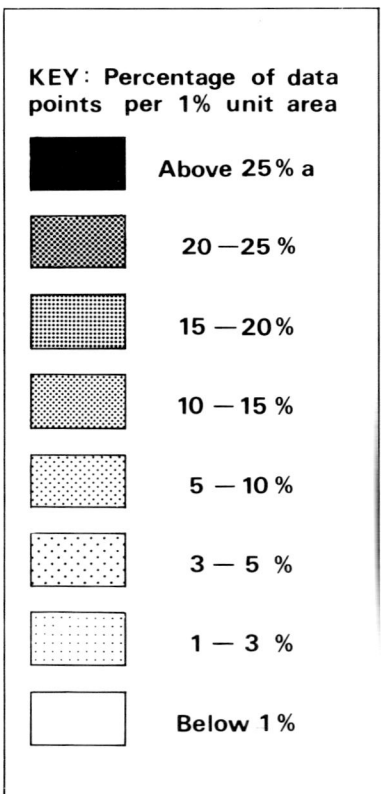

KEY: Percentage of data points per 1% unit area

Above 25% a

20 — 25 %

15 — 20 %

10 — 15 %

5 — 10 %

3 — 5 %

1 — 3 %

Below 1 %

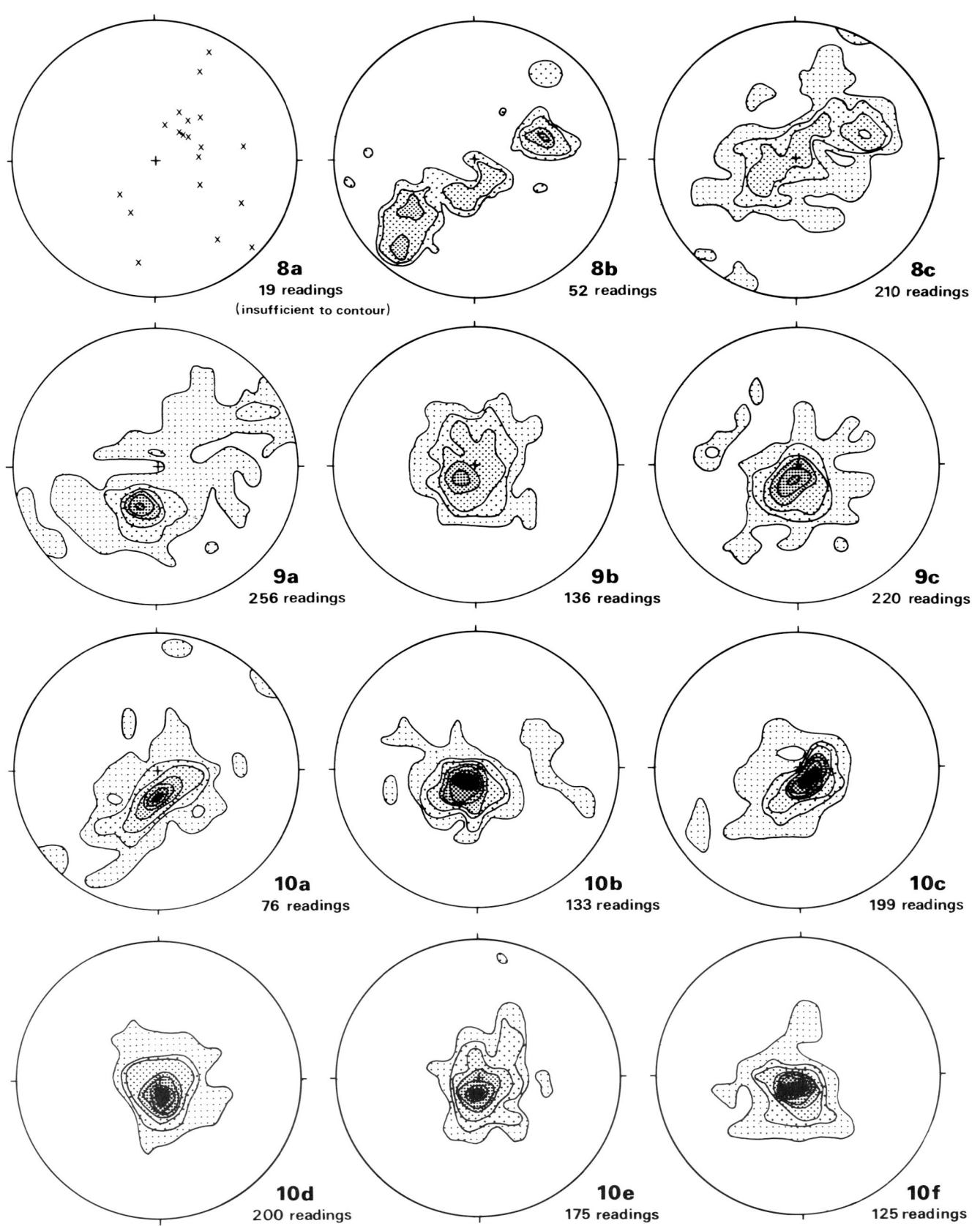

8a
19 readings
(insufficient to contour)

8b
52 readings

8c
210 readings

9a
256 readings

9b
136 readings

9c
220 readings

10a
76 readings

10b
133 readings

10c
199 readings

10d
200 readings

10e
175 readings

10f
125 readings

The sub-areas have been delineated to give some measure of stratigraphic control to the presentation of the data:

8A The Suta Volcanics.

8B The upper Itina Valley, corresponding to the lower part of the Kavo Greywacke Beds and intercalated Suta Volcanics.

8C The middle Itina Basin (the diagram for this area was constructed from field data supplied by Dr. R. B. Thompson).

9A Tina Zone—corresponding to the eastern two-thirds of the combined outcrop of the Mbetilonga Limestone, Tina Calcarenite and Tangareso Beds, i.e. corresponding broadly with the Lower Miocene.

9B Matepono Area—including the Gold Ridge Volcanics and surrounding Toni Formation.

9C Kombusoe Area—mostly Toni Formation.

10A The Lake Lee Calcarenite outcrop.

10B The Valasi–Longgu Zone, corresponding to the outcrop of the Valasi Limestone and Longgu Beds.

10C The Lower Manuhoho Area, corresponding to the western half of the outcrop of the Mbokokimbo Formation.

10D The Kalangali Area, the central part of the outcrop of the Mbokokimbo Formation.

10E The Paripao–Reko Area, based on the Mbalanga Shale and intercalated conglomerates.

10F The North-East (Rere–Aola) Zone, corresponding broadly with the outcrop of the Vatumbulu Beds.

According to Coleman (1965a) the sediments in Guadalcanal are draped over the faulted basement, thickening to the north-east to give a wedge-shaped cross-section. Referring to north-central Guadalcanal in particular, Coleman (1960) states that 'dip values (of the sediments) tend to decrease up-section from 20° to 25° for the Miocene basal beds to 5° to 10° for part of the Mt Austen Beds'. Although aware that in detail the structure of the sediments was more complex than this, Coleman considered that there was no folding 'apart from minor undulations'.

The contoured stereograms show that as far as north-central Guadalcanal is concerned, Coleman's impression was substantially correct.

The decrease in dip values up-section is confirmed by the maxima obtained from each sub-area:

9A	Lower Miocene to Middle Miocene	25° NNE
10A	Lower Miocene	10° NW
10B	Lower Pliocene	9° NE
10C	Upper Miocene(?) to Lower Pliocene	10° NW
10D	Upper Miocene(?) to Lower Pliocene	8° N
9B	Mainly Pliocene	11° NE
9C	Mainly Pliocene	9° NNE
10E	Mainly Pliocene	8° N
10F	Mainly Pliocene	6° NNE

There is a better-defined pattern with a broader range in amount and greater variation in direction of dip in the Lower and Middle Miocene than in the Pliocene, although the effect is somewhat obscured by the rather wide scatter of points in the Pliocene; part of this scatter is probably an expression of the greater initial irregularity in the stratification of the conglomerates, and consequent difficulty of making accurate measurements. Some variation in the direction and amount of dip may also be attributed to differential compaction above an irregular interface at the base of the sedimentary pile.

The presence of an open synclinal structure in the Lower Miocene sediments was suspected from examination of the aerial photographs between the upper Kolombolavu and Mongga Rivers. In this connection stereographic plots for sub-areas 10A and 10C show the same kind of distribution; although the maxima in both cases indicate a predominantly low angle of dip to the north-west, there is a great circle distribution corresponding to rotation about an axis plunging at 10° to 15° to the north-west.

In the case of sub-area 9A, there is a different symmetry; in the Mbetilonga Limestone and Tina Calcarenite there are two maxima: the 'regional' dip of 25° to the north-north-east, and a secondary maximum, around 70° to the south-west. Both fall on a great circle plunging at a low angle to the north-west; this effect could be regarded as due either to rotation about a swarm of closely spaced faults trending north-west, or to monoclinal folding.

The writer has delineated a series of north-easterly-trending fold axes in the headwaters of the Tinahulu River, no fold turnovers have been identified in the field; although a few faults with north-easterly trend have been mapped, there are no parallel lineaments on the aerial photographs.

The Lower Miocene in the Upper Tinahulu area is interpreted as having been involved in a series of monoclinal folds, the steep limbs consistently facing south-west. It may fairly be assumed that the structure in the Lower Miocene is a response to faulting in the underlying pre-Miocene basement.

Disposition of the sediments in western Guadalcanal

Some detailed structural information on post-Oligocene sediments is available for the Itina Basin (Thompson, 1968) and the Lungga Basin (Wright, 1968b). In the Itina River Valley the sediments tend to dip at fairly steep angles either to the south-west or north-east. The inference is that there has been folding about axes trending north-west to south-east (see stereographic projections for sub-areas 8B and 8C, Figure 13).

From Wright's work in the Lungga Beds to the north-west, it is clear that there is a low regional dip of 5°–30° to the north or north-north-west, with local variations due to faulting.

Pudsey-Dawson and Thompson (1958) showed that the Honiara Beds dip at low angles, 5°–15° to the north or north-east.

Steeper dips and more variable strike trends have been recorded from the outcrop of the Mbonehe Limestone. In the extreme west of the island Pudsey-Dawson and Thompson (1958) recognise an east–west 'trough' affecting the mudstones of the Lungga Beds in the Tangarare and Hoilava rivers; from the detail on their accompanying map the axis of folding appears to trend north-west or west-north-west, i.e. parallel to a prominent fault swarm to the south of the area, which also reflects the linear trend of the Ghausava Ultrabasics.

In the Mbetilonga area Hill (1960) regarded the Mbetilonga Limestone, Tina Calcarenite and Tangareso Beds as involved in a highly faulted synclinal structure; the axis of the fold was presumed to trend east–west, although this is not clarified in the published map and sections.

Wright (1968b) drew stratum contours for the base of the Mbetilonga Limestone in the same area; using the contoured 1:50 000 AMS sheet No. 7729 II he concluded that, apart from minor irregularities, the configuration of the outcrops in the Mbetilonga area could be accounted for if the upper surface of the pre-Miocene basement rocks were a peneplain dipping at about $2\frac{1}{2}°$ towards the north.

These impressions of the structure of western Guadalcanal confirm the pattern of more variable dips, locally indicating broad folding on north-westerly axes in the older half of the sedimentary sequence, and low angles of northerly dip in the younger formations; the degree of disturbance due to faulting decreases up-section.

Inferences drawn from the examination of a few sections could be very misleading; in spite of the effects of faulting the presentation of data in the form of density diagrams for a given area appears to give a fair impression of the regional homogeneity underlying the 'scatter'.

Folding

Minor folds within the Mbirao Group

In spite of the widespread development of schistosity within the Mbirao Metabasics, only 65 mesoscopic folds have been recorded in the field over the entire area of the Mbirao Group. Original stratification, defined by thin bands of pelagic limestone, has been observed in only 13 of these examples; in the other cases the folding has affected metamorphic s-surfaces, predominantly S_1. Three-dimensional access is not available in many instances; however the plunges of 26 of these folds have been plotted on a stereogram (Figure 14A). Fourteen of the readings form a distinct group plunging at 10° to 40° towards the north-west: these have been designated as primary F_3–F_4 folds, since their axial planes are parallel to S_3 and S_4 surfaces.

The wavelength of these minor folds ranges from a few centimetres to 10 m; their styles vary from highly angular 'chevron' shapes to concentric warps of low amplitude. They may be classified tentatively according to the relationships with s-surfaces within the metabasics:

1   F_1 folds are minor intrafolial folds in the sense of Turner and Weiss (1963, p. 116). They are essentially tight monoclinal flexures with axial planes parallel to S_1: accordingly they are assumed to have been formed during the first phase of tectonism. Good examples occur in the upper Alivaghato Valley.

2   F_2 folds have their axial planes parallel to S_2. Isolated examples occur in the Na Humbu (Salinaho) tributary of the Kolovaghamela River, in the Na Papa tributary of the Kolohaisava and the Haimela section of the Nggeunaha River. At the first two localities the fold turnovers are defined by stratification in the Tetekanji Limestones: at Haimela they are defined by S_2 surfaces.

3   Primary F_3–F_4 folds have their axial planes parallel to the strike of S_3 or S_4, and often have a well developed axial plane foliation; 40 of the 65 observed folds fall within this category. Swarms of these folds have been mapped in the Charikiki tributary of the Kolohaisava and also in the upper

Rere River; at the former locality the Tetekanji Limestones have been deformed in a rather irregular 'flow-fold' style (Turner and Weiss, 1963, p. 481) although a secondary axial plane foliation has developed in the intercalated chlorite schists. In the upper Rere section the folds are open, of low amplitude, but with quite angular turnovers; S_1 surfaces have been warped in a similar manner in the upper part of the Kolombolavu River. All these folds plunge consistently at low angles to the north-west.

Figure 14 Stereograms showing the plunge of fold axes in Guadalcanal

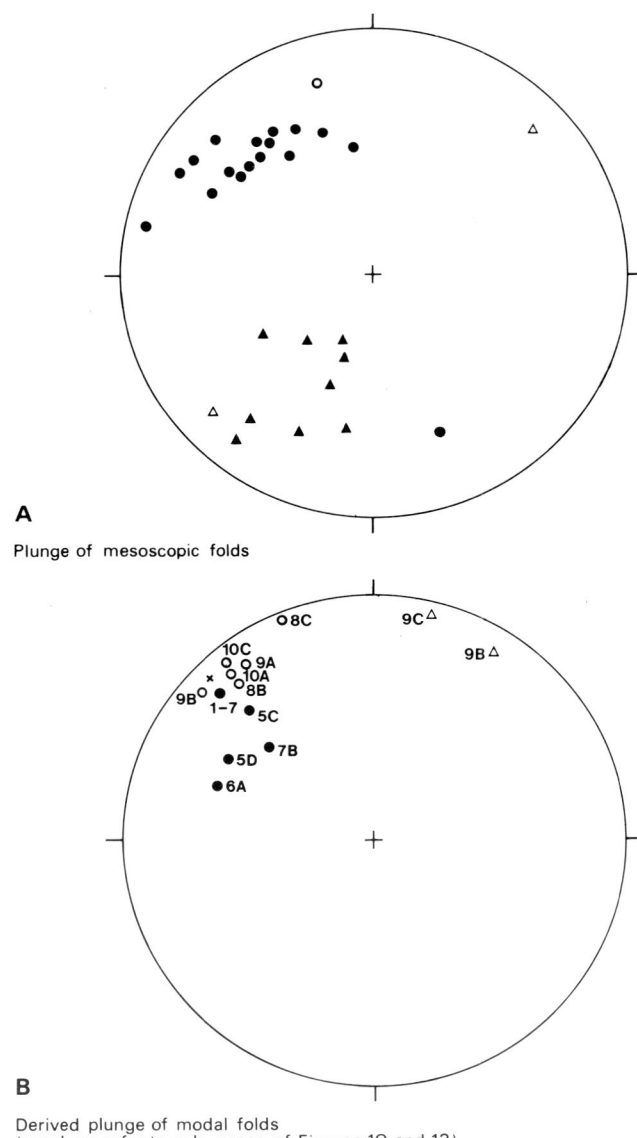

A

Plunge of mesoscopic folds

B

Derived plunge of modal folds
(numbers refer to sub–areas of Figures 10 and 13)

● Primary F3/F4 folds in Mbirao Group

○ Primary F3/F4 folds in post–Oligocene sediments

▲ Secondary F3/F4 folds in Mbirao Group

△ Secondary F3/F4 folds in post–Oligocene sediments

× Modal fold, Kwara'ae anticline, north Malaita

4   Secondary F_3–F_4 folds. Minor folds have been recorded, mainly from the eastern half of the Mbirao Metabasics belt, with axial planes trending west-south-west. In the upper Na Ndoli River one such fold affects quartz veins parallel to S_1, and is in turn transected by S_4 surfaces; possibly these folds are related to the S_3 tectonic phase, but were not orientated normally to the maximum regional horizontal stress; such instances could be interpreted simply as drag responses to local stresses related to faulting. Secondary F_4 folds occur in the western part of the Mbirao Metabasics outcrop, notably the Haimela section of the Nggeunaha River. Their axes plunge south-west or south-south-west. (Figure 14A), and as these folds post-date S_1, S_2 and S_3, they are here regarded as belonging to a late, possibly Pliocene, phase of tectonism. The fold styles vary from concentric to highly angular. Other incidental folds may be of comparatively recent 'drag' origin, for example folding of S_2 surfaces in the Mbolavu River, in close proximity to the Veeru Moli Fault.

Minor folds within the Neogene succession

Only eight mesoscopic folds have been mapped within the post-Oligocene sediments: the axes of three of these trend north-west, the other five north-north-east. They are all low-amplitude concentric-style flexures, and cannot be traced for any great distance along the strike. The pattern is therefore essentially one of incidental, sporadic folding; some of the effects are rather impersistent, or disharmonic, and therefore probably expressions of pre-consolidation movements or differential compaction on an irregular substratum under a variable load. However, the consistency in trend, even of these minor isolated structures, suggests that there may be some form of patterned regional control. In this connection Wright (1968a) ascribes a north–south anticlinal flexure in the Toni Formation exposed in the middle Tina River to dextral transcurrent movement along a buried basement fault trending north-east.

A revealing section in the Chovohio River between Tinomeat and Charihaivati shows a faulted sequence of monoclinal folds whose axes trend north-north-east. These structures affect the Upper Pliocene sediments and are parallel to the axes of the anticlinal horsts of the Tausoro and middle Tina valleys.

Occasionally small mesoscopic folds can be clearly related to movement on adjacent fault planes.

Regional folding

Allum (1967) drew attention to evidence for gentle folding in the Mbokokimbo Beds of east-central Guadalcanal, pointing out that all the folds appeared to be synclinal. Two synclinal axes trend north-west in this area, the Haiparia and the Chimiu synclines: possibly a third syncline crosses the head of the Aola River. The structure of the upper Tinahulu River and the Itina Basin has also been interpreted in terms of folding on north-westerly axes.

In the absence of continuous mesoscopic folding the geometry of the flexures can only be surmised from statistical diagrams and the construction of cross-sections. Figure 14B

is a synoptic plot of the plunges of modal folds (Williams, 1959) corresponding to the girdles derived from Figures 10 and 13; only well-defined girdles are represented originating from a total of 1817 s-surface measurements. The modal folds plunge consistently towards the north-west whether derived from the bedding planes S_0 in the Mbirao Group, metamorphic s-surfaces in the Mbirao Metabasics or bedding planes in the Lower Miocene. Fold axes in the sediments tend to be orientated more towards the north-north west, those in the metabasics to the north-west.

From the relationships with the regional trend of S_3 and S_4 surfaces, it may be inferred that gentle folding of the basement schists on north-westerly axes was associated with the imprinting of S_3, whereas S_4 formed in the Mbirao Group after the deposition of the Lower Miocene sediments, which adjusted to shearing and faulting in the basement in part by gentle folding. However the anticlinal crests were 'faulted out', for example along the line of the Kolombolavu Fault between the Haiparia and Chimiu synclines in east-central Guadalcanal. In the Itina River area, however, anticlinal crest faulting does not appear to have affected the Kavo Greywacke Beds.

Minor folds trending west-north-west also affect the outcrop of the Lake Lee Calcarenite in the Manuhoho River section. These could be regarded either as having formed during the F_3 phase, or as a secondary response to local sinistral wrench-faulting in the basement during the F_4 phase.

Later (Upper Pliocene) folding appears to have been limited to north-central Guadalcanal, where impersistent 'fault folds' with axes trending north-north-east give rise to local anticlinal horsting features, for example the Tina flexure and the Tausoro anticlinal horst. The faults and minor folds of comparable trend occurring in the same area are here considered to be related to the same phase of tectonism.

Faulting

Fault lineaments

Several authors have emphasised the importance of large-scale faulting in the Central Province of the Solomon Islands.

According to Coleman (1965a) most of the major faults in Guadalcanal trend north-west and west-south-west, giving rise to a trellis-work pattern. 'They are high-angle with predominant dip-slip components, but some have strong strike-slip components as well.'

Information about the pattern of faulting has been obtained in two different ways:

1   From a consideration of the major lineaments as recorded from aerial photographs, noting in particular any evidence of displacement of Recent geomorphologic features.

2   From the compilation of trend diagrams based on the attitudes of fault planes actually measured in the field.

Allum (1967), with reference to the photogeological interpretation of Choiseul, remarks that 'the number of lineaments probably caused by faulting is enormous; under such circumstances, the number of faults interpreted is roughly proportional to the time spent on the interpretation.' This would be equally applicable to the photointerpretation of the Mbirao Group and the Gallego Volcanics.

Major faults

From general geologic considerations, there are within Guadalcanal no structures which might be designated as 'master faults' in the sense of primary fractures of considerable vertical or horizontal displacement, to which subsidiary structures could be genetically related.

There are two possible exceptions to this generalisation:
1   The line of ultrabasic bodies, which demarcates, in the Ghausava, Suta and Marau areas, the boundary between the Guadalcanal Gabbro and the Mbirao Volcanics (or their metamorphosed equivalents).
2   The Suta Fault, which brings the Mbirao Metabasics and the Guadalcanal Gabbro into juxtaposition with the Suta Volcanics; the line of this fault follows the northern boundary of the Kavo Ranges, but its north-western termination is obscured on the southern margin of the Lungga Basin—it is uncertain whether or not the Ghausava Fault Zone could be regarded as its western continuation.

There is no direct evidence that either of these structures has been active since the Upper Miocene.

Faults trending east-south-east are clearly of importance in the eastern half of the Mbirao Block.

The Valasi Fault Zone, comprising two or more anastomosing fracture systems, can be traced over a distance of 100 km; at its western end it separates the Valasi Block from the Mbirao Metabasics, and may be considered to be part of the fracture system into which the ultrabasic bodies have been intruded. Chlorite-actinolite-schists of the Mbirao Metabasics are in juxtaposition with relatively unaltered lavas along this line; it may be inferred that there has been a considerable downthrow to the north, assuming that the metamorphism is in part a function of depth, and that the schists have now been denuded in the axial region of a composite anticlinal horst. However, the relationships between the Guadalcanal Gabbro and the Mbirao Metabasics in the Suta and Marau areas suggest that the ultrabasics were emplaced into a high-angle thrust zone, the Guadalcanal Gabbro having been thrust over the metavolcanics towards the south-east in Suta, and towards the south in the Marau area. Later downfaulting to the north, which in part incorporated the fracture system in the ultrabasics, brought about the preservation of a plate of unaltered Mbirao Lavas which became the Valasi Block. Renewed faulting along these lines also affected the Lower Pliocene limestones in the gorge of the middle Vaghanambo River.

The Tetekanji and Mandonu Faults are parallel to the Valasi Zone, forming part of the boundary between the Mbirao Metabasics and the Weather Coast Block; they could be interpreted as normal faults with considerable downthrow to the south.

Faults of the same trend are prominent in a zone fringing the south coast from Cape Henslow westward to the Kuma River; it is clear from an examination of the high-point strip profiles (Figure 6) that these faults form a sequence of steps, an expression of downfaulting to the south which has controlled the position and shape of the present-day south-ern coastline. The steps are also apparent in the river profiles.

The Chimiu Fault Zone trends in a south-easterly direction for about 25 km from Lee's Lake in the upper Mongga Valley, following in part the line of the upper Sambahalava River. It terminates abruptly 2 km north-west of Lee's Lake: the overlying Lower Miocene sediments have not been displaced; parallel minor structures include the Hanangga and Oa Faults to the east. All three faults are prominent as lineaments, but their net displacement does not appear to have been very considerable.

The Suta Fault, which forms part of the western boundary of the Mbirao Block, also trends north-west; it may be traced for at least 50 km to the north-west from the mouth of the Riva River. The Suta Fault must have a downthrow to the south-west of several hundred metres; it is also in part a high-angle reverse fault; from the relationships in the upper Kuma Valley, where the Mbirao Metabasics and Suta Volcanics are in juxtaposition, on topographical grounds alone a relative movement of over 1000 m would be required.

Faults trending north-east are prominent in central Guadalcanal; the Malango Fault appears as a distinct lineament on the aerial photographs to the north of Nduindui village: it traverses very high ground and the relative movement is consequently difficult to interpret, but from the displacement of boundaries in the Tina headwaters the impression is that the basement and Lower Miocene rocks have been displaced 1 to 2 km sinistrally, a vertical component implying downthrow to the north-west. Possibly this operated as a sinistral oblique-slip fault in the Middle to Upper Miocene; traced to the north-east of Gold Ridge it does not appear to have displaced the Pliocene conglomerates, although its presence in depth is suggested by the rectilinear middle course of the Tinahulu River; to the north the trace is lost in a complex of minor north-west-trending left-lateral faults in the lower part of the same river. The Mbalisuna Fault Zone is a parallel system reflecting post-Pliocene movement along a belt of crustal weakness collinear with the trend of the Suta Ultrabasics.

The overall trends of faults whose traces exceed 10 km in length, or which form a prominent system of parallel fractures, are shown on the accompanying trend diagram (Figure 15). These major lineaments tend to strike west-north-west, north-west or north-east. They might be considered in a general way to parallel the main trends of s-surfaces in the Mbirao Group, leading to the suggestion that in the first instance they were initiated as normal or thrust faults parallel to axes of anticlinal horsting; subsequently these fractures were incorporated into a succession of different stress systems, sustaining a sequence of diverse roles which involved dip displacements locally antithetic to the original motion.

Strike-slip faulting

The sequence of s-surfaces within the Mbirao Group suggests that the principal horizontal stress may have changed in direction, in relation to the present-day north, from north-west to south-east in the earliest Tertiary(?) to north-east to south-west in the late Miocene. During three or possibly more separate periods in the tectonic history of

Figure 15 Diagrams showing the trends of major faults in Guadalcanal

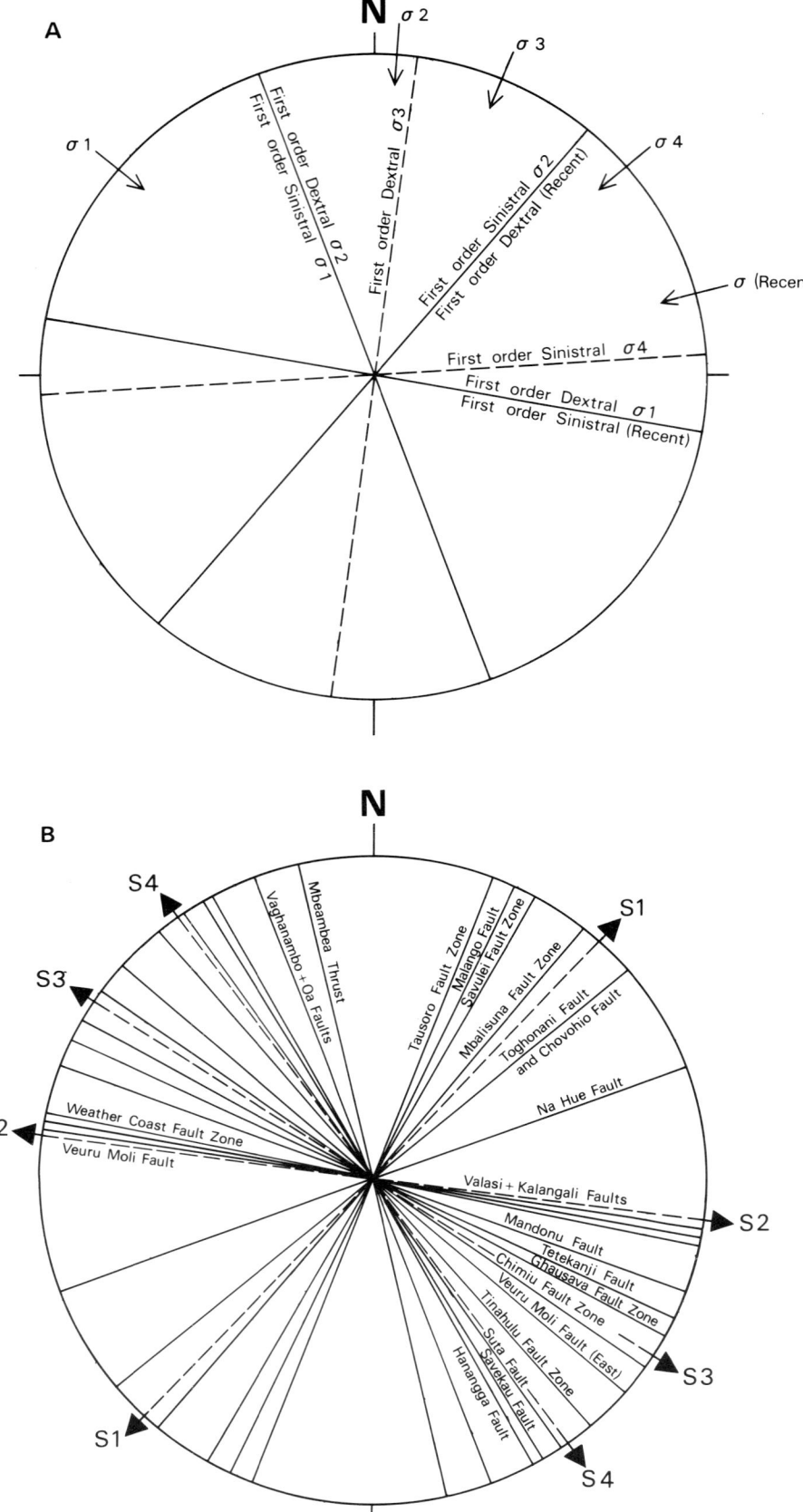

the Mbirao Group the regional stress effectively imprinted a distinct set of s-surfaces, each with fairly consistent orientation. Each of these tectonic events might have imposed its own related systems of folding, conjugate shears, and joints on the complex as a whole: in the later sequences, pre-existing structures might be incorporated into the dynamic pattern, but would alter their function and status in sympathy with the change in orientation of stress.

It is important in this context to consider the possible role of transcurrent faulting. Thompson (1960) pointed out that the outcrop of the Marau Ultrabasics was displaced at the eastern end by dextral shear along faults trending north-west. Coleman (1960a) tentatively interpreted the fault pattern on Choiseul as having arisen from major crustal shearing in a roughly east–west direction. However, references to evidence for transcurrent movement do not commonly feature in the records of field workers in the Solomons, rather the recurrent theme has been the importance of vertical movement and block-faulting.

Evidence for transcurrent movement

Indications of strike-slip components were interpreted from the following phenomena:

1 Offsetting of drainage patterns or other geomorphologic features.
2 Displacement of planar structures in the country rock, such as dykes, foliation, s-surfaces.
3 Zones of tension cracks or feather joints indicating sense of movement on an adjacent fracture.
4 Drag in s-surfaces adjoining fault planes.
5 Slickensides on fault planes.

The first two criteria seem to be more reliable than the last three, inasmuch as small-scale structures associated with particular faults, such as slickensides, may indicate minor recent dislocations whose sense is opposed to that of an earlier protracted history of movement. Offsets of geomorphologic features can be regarded as relatively recent effects, and displacement of particular s-surfaces can often be fitted into a particular historical sequence: however secondary small-scale structures cannot always be fitted unequivocally into the movement picture.

Figure 16 shows how the course of the Sambaharihi River has been deflected along the line of the Veuru Moli Fault, the most conspicuous fracture in the Weather Coast Fault Zone. It might be inferred that there has been a sinistral displacement of 1 to 2 km along west-north-west lines since the establishment of the Sambaharihi valley system.

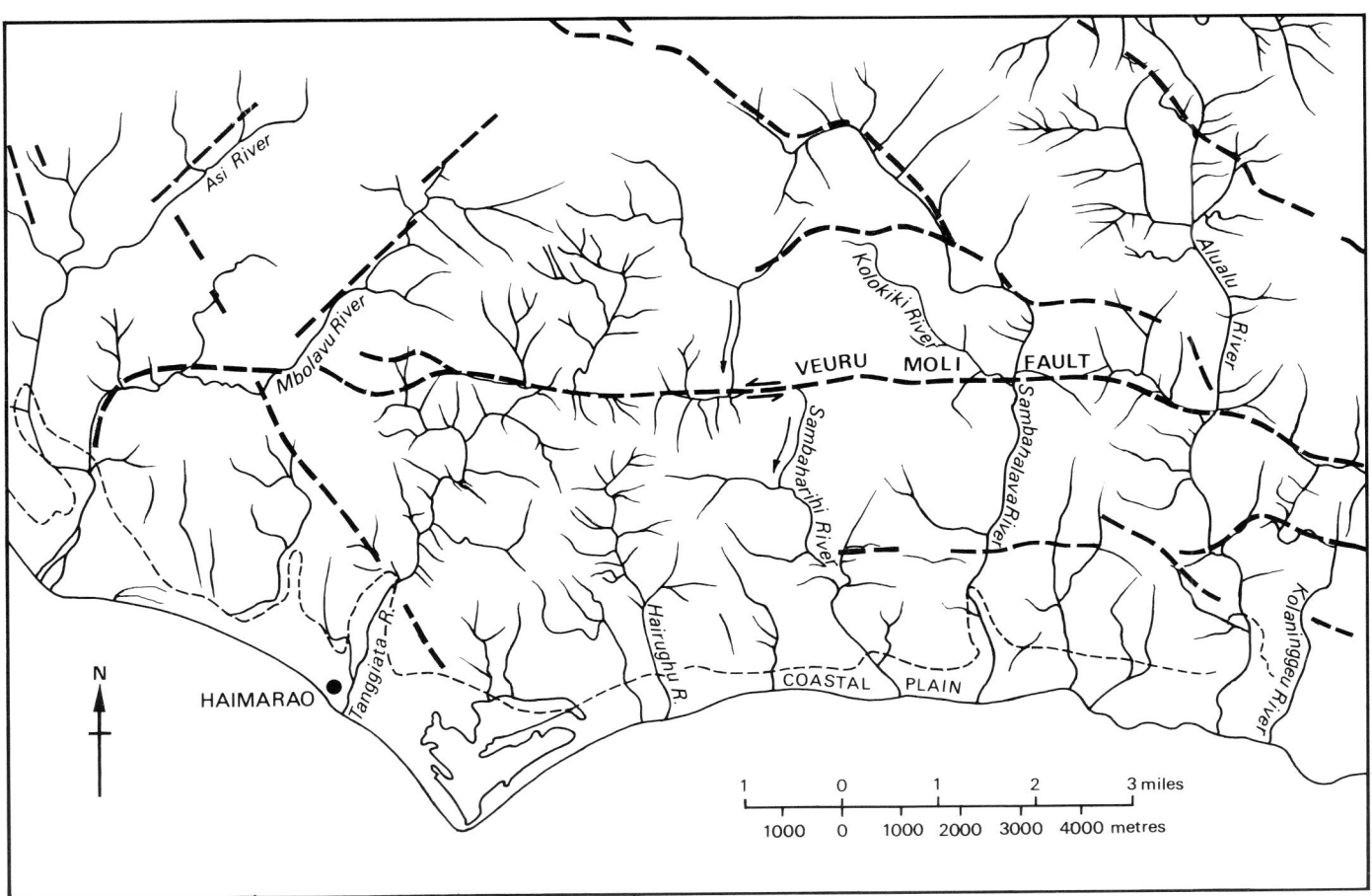

Figure 16 Map showing the influence of the Veuru Moli Fault on the pattern of drainage in south-eastern Guadalcanal

This effect is most obvious in the case of the Sambaharihi, but where the line of the Veuru Moli fault crosses the Tanggiata River to the west and the Sambahalava, Alualu and Kolaninggeu rivers to the east, it is clear that a pattern of 'shutter ridges' has formed to the north; in each case the drainage has been deflected along the strike of the fault.

Coleman (1960b) remarked on the swarm of strike-slip faults in the lower Tinahulu River. These give rise to a series of fresh scarp faces in the Pliocene conglomerates and sandstones; the displacement is consistently sinistral on north-west lines but does not amount to more than 500 m on any one fault.

The attitudes of 877 fault planes have been measured throughout central and eastern Guadalcanal. Of these, 83 show some evidence of transcurrent movement: 47 are interpreted as sinistral and 36 as dextral. Figure 17 is a composite rose-diagram for wrench faulting; a maximum of faults with left-lateral movement trends west-north-west, and a subsidiary swarm with predominantly right-lateral movement trends east-north-east. The other maxima are equivocal, indicating both sinistral and dextral movement along the same trend. Eight of the west-north-west sinistral faults are interpreted as having been active during the Holocene, since the sense of movement has been derived from the offsetting of geomorphologic features. Three of the faults, two of which are dextral trending east-north-east, affect the Neogene sediments, so that the majority are from the Mbirao Group and might be related to one of three or four different phases of tectonism.

Figure 17 Rose-diagram showing the trends and character of transcurrent faults in Guadalcanal

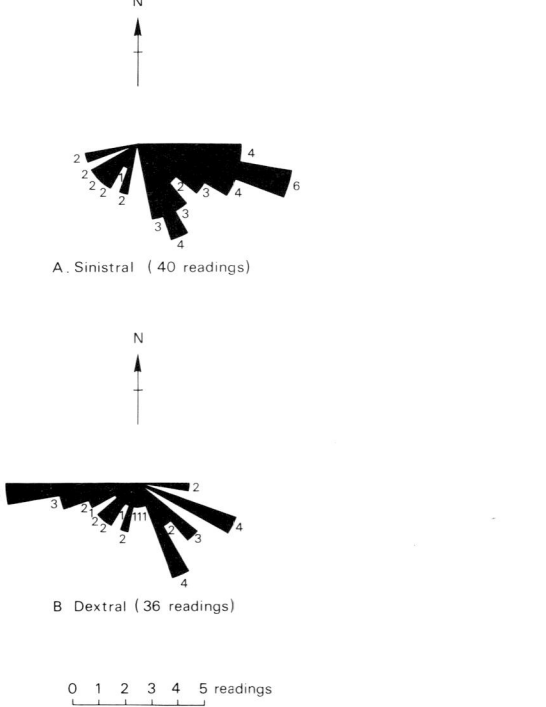

A. Sinistral (40 readings)

B. Dextral (36 readings)

0 1 2 3 4 5 readings

Interpretation of fault patterns

Moody and Hill (1956), in part incorporating the ideas of McKinstry (1953), developed a system for analysing patterns of strike-slip faults. The concept has been examined by Hills (1963, p. 207) and discussed in some detail by Badgley (1965, pp. 261–272). Based on these concepts, a hypothetical composite fault pattern was devised for comparison with the patterns derived from field observations (Figure 15A). This pattern incorporates only the first-order conjugate shears in relation to assumed directions of maximum horizontal stress, as derived from the attitude of s-surfaces and fold axes. Recent sinistral east-south-east faults are also included in the pattern together with their theoretical dextral counterpart.

Perusal of the diagram shows that by virtue of their orientation earlier fractures are likely to be incorporated into later patterns with a reversed sense of movement as the orientation of stress changes.

The following are the major correspondences:
Sinistral for Phase 2 with Dextral Recent Phase ~ 40°
Dextral for Phase 1 with Sinistral Recent Phase ~ 100°
Sinistral for Phase 1 with Dextral Phase 2 ~ 160°
The composite diagram for transcurrent faulting (Figure 17) shows the following maxima:
1 Predominantly Sinistral ~ 100°
2 Predominantly Dextral ~ 80°
3 Dextral-Sinistral ~ 160°
4 Dextral-Sinistral ~ 40°
5 Dextral-Sinistral ~ 10°
The maxima at 160° and 40° correspond in kind and direction with two of the theoretical 'reincorporated' directions. Although the first-phase dextral component is not in evidence around 100°, the maxima at 10° and 80° also correspond with theoretical first-order shear directions.

It may be concluded that the field data correlate well with hypothetical correspondences between first-order shears as predicted by the Moody and Hill system.

Faulting in the Mbirao Group and Suta Volcanics

Figure 18 shows strike frequency diagrams for fault planes actually observed in the field. The dip of the fault planes does not deviate more than 20° from the vertical for 90 per cent of the readings.

In area 5 (west Mbirao) there is a very good correspondence between the maxima predicted theoretically on Figure 15 and those derived from the mesoscopic data.

In the other areas at least three of the maxima correspond with maxima on the composite diagram; however, in the Suta–Koloula division (area 8) the pattern shows less conformity and may be more usefully compared with the systems of jointing and mineralised veins (Figures 19 and 21): this suggests that in that area the faulting reflects the tensional effects associated with emplacement of the Koloula Diorite, rather than the regional first-order shear pattern predicted from the Moody and Hill system.

Throughout the Mbirao Block, but particularly in areas 2, 6 and 7, the importance of the west-north-west fault trend, parallel to the Weather Coast and Valasi fault zones, is most evident. The transverse high-point strip profiles

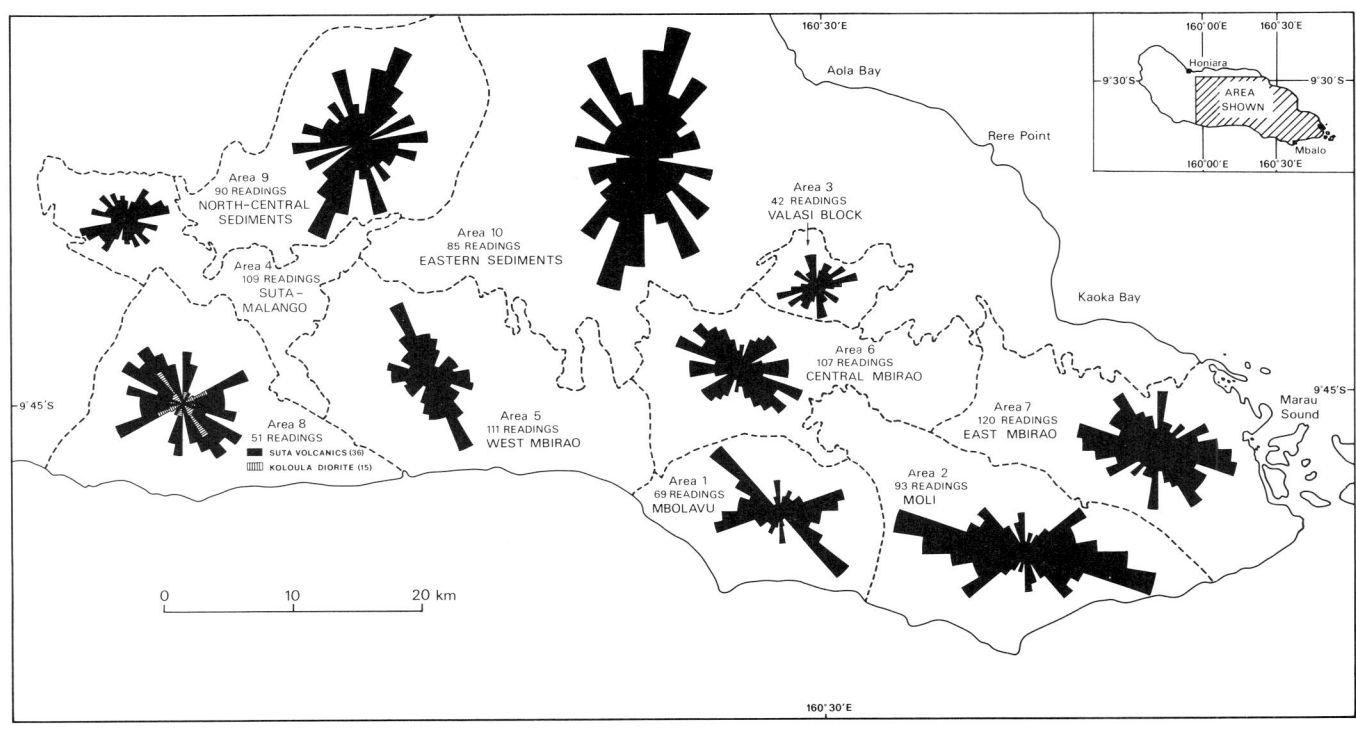

Figure 18 Rose-diagrams showing the strike of faults in eastern and central Guadalcanal

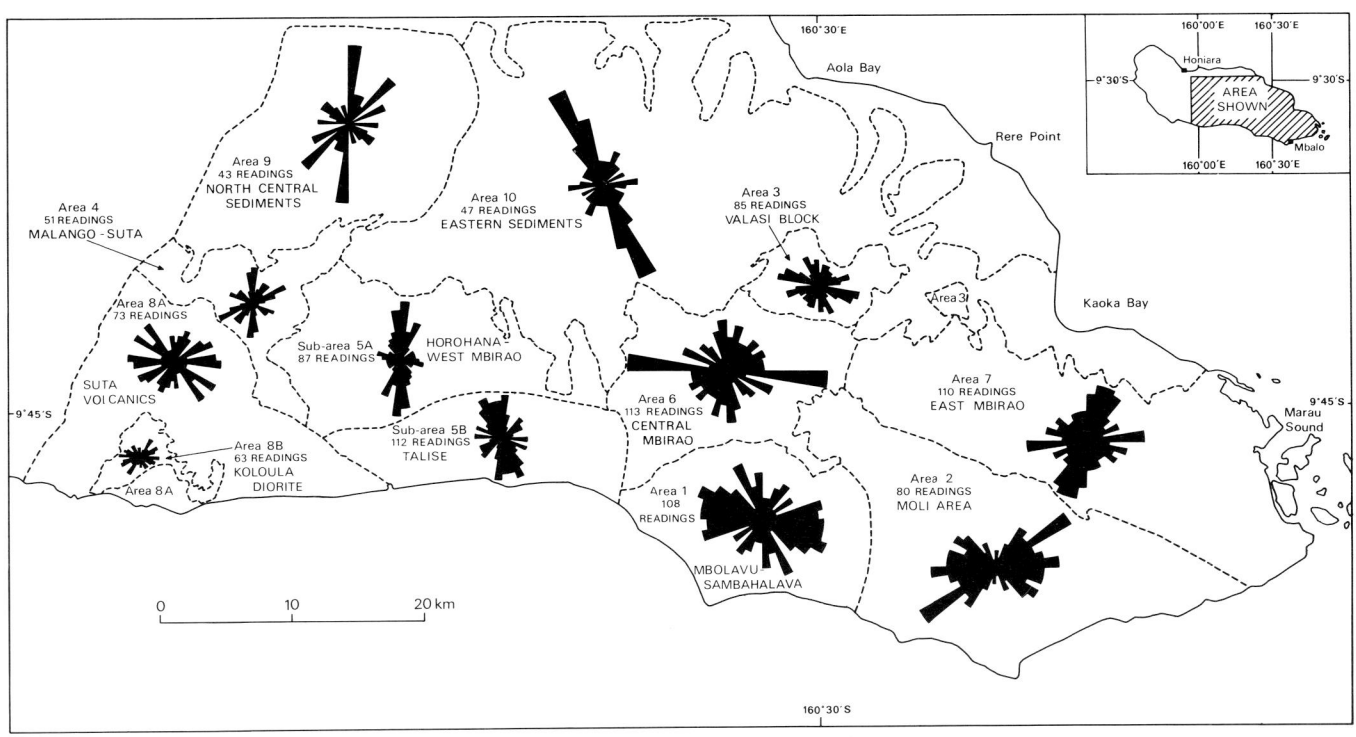

Figure 19 Rose-diagrams showing the strike of joints in eastern and central Guadalcanal

(Figure 6) display evidence for progressive downthrow along major step-faults of this trend.

In the western half of the Mbirao Block the dominant fault trend swings round to the north-north-west, parallel to the S_4 surfaces which are so well developed in the river valleys between the Riva and Alivaghato (Figure 10, sub-area 5F).

At the head of the Mborambora tributary of the Mboko-kimbo River, Mbirao Metabasics have locally been thrust eastwards over Lower Miocene Lake Lee Calcarenite; the strike of this Mborambora Thrust is north-north-east, parallel to the main trend of the Suta Ultrabasics and to lines of Pliocene faulting in north-central Guadalcanal; the thrust plane dips at 50° to the west. The direction of movement accords with a general uplift of the centre of the island along the line of the Suta–Horohana 'axis'; the longitudinal high-point strip profiles also indicate progressive step-faulting downwards to the east and west of the Kavo–Horohana 'high'.

Faulting in the sediments

North-central Guadalcanal (area 9), west of the Sutakama Valley, has been much more extensively faulted than the area of sediments to the east. Thus in Figure 18 the rose-diagram for faults actually measured in the field in area 10 is based on about the same number of readings as for area 9, although the eastern area is over four times as large.

However, there is good correspondence between the fault patterns in the two areas. In both cases there is a distinct maximum striking north-north-east, and a secondary maximum trending north-north-west. Faults of both trends have clearly displaced the Lower Miocene limestones both east and west of the Sutakama Valley.

Relationships in the Tausoro Valley show that faults trending north-north-east have played a part in minor anti-clinal horsting where the Tausoro River has cut into the crest of a faulted dome to expose a small inlier of the Tina Calcarenite within the surrounding Pliocene conglomerates.

In the Lame River section, conglomerates of the Toni Formation have been sheared in zones up to 2 m wide trending north-east; the phenoclasts have been deformed into ellipsoidal shapes whose long axes are horizontal, but orientated at an angle of about 40° to the margins of the fault zone.

Faults trending north-north-west are in part penecontemporaneous in the Mbalanga Shale immediately overlying the Toni conglomerates in the Mberande River section; this gives a probable Upper Pliocene date to this phase of faulting, which reflects trends in the underlying basement, and also parallels the axes of folds in the Lower Miocene sediments.

The pattern of faulting in western Guadalcanal

Fault-plane data for readings taken in the field are not available for western Guadalcanal. Pudsey-Dawson and Thompson (1958) considered jointing and faulting in that area were related to at least two periods of earth movement.

Photogeological interpretation reveals the presence of two important fault complexes:

1 The Savulei Fault Zone, a swarm of faults trending north-east cutting across the north-west corner of Guadalcanal; these faults are associated particularly with the outcrop of the Gallego Lavas: locally they bound inliers or horst-like slivers of the underlying basaltic andesites of 'Suta Volcanics' type. Faults of this trend, parallel to the Malango Fault, are prominent in the sediments of north-central and east Guadalcanal. The quiescent Recent volcano of Savo Island is transected by a prominent fracture trending north-east, the locus of most of the active thermal areas in the island; possibly this represents the north-easterly extension of the Savulei Fault Zone.

2 The Ghausava Fault Zone, a swarm trending west-north-west, is conspicuous on the aerial photographs on the north side of the Ghausava Ultrabasics, affecting both the Mbirao Volcanics and the Lungga Beds. There is a suggestion of an 'en échelon' zone linking the Ghausava Fault Complex with the Suta Fault to the east.

The dominant trends of these two complexes could be regarded as corresponding to a pair of conjugate shears related to a maximum horizontal regional stress oriented in a north-east direction. Other prominent lineaments also correspond with the theoretical composite fault pattern (Figure 15A), for example the Itina Fault which trends north-north-west across the lower Itina Valley and forms the western boundary of the Itina Ultrabasic body; it may be traced as a discontinuous lineament as far north as the Mavo River.

The overall faulting pattern in western Guadalcanal is broadly comparable with that of the rest of the island; there are no master faults, rather movement appears to have been concentrated locally in complex zones of intense fracturing.

The general impression gained from the geological map is that the Pliocene sediments have been much more intensely faulted in north-west Guadalcanal and the Gold Ridge area of central Guadalcanal than elsewhere; correspondence with areas of Plio-Pleistocene vulcanicity is thus apparent.

Jointing

The trends of 972 joint planes for eastern and central Guadalcanal are incorporated in the rose diagrams in Figure 19. These are essentially joints in the sense of Billings (1942, p. 106), i.e. 'divisional planes or surfaces that divide rocks, and along which there has been no visible movement parallel to the plane or surface'. The concept of shear joints, in contradistinction to tension joints, connotes a tendency towards differential movement, as in a pair of conjugate shears; although no visible evidence for movement may be detected, the mechanism requires that some such relative motion is likely to have occurred in many instances. In the Mbirao Group sets of closely spaced joints have frequently been recorded, for which no evidence of movement is demonstrable, yet they may parallel neighbouring shear zones where there has clearly been some differential motion. Particularly in such a domain of polyphase deformation, where pre-existing structures become incorporated in changing movement patterns, the sequence joint–shear joint–fault becomes a continuum, and the mapping of a given structure in one or another category depends on subjective considerations.

Masashige (1969) divides joint-patterns into two major types: Parallel, with a single peak in the strike-frequency diagram; and diagonal, with two peaks in the diagram.

In terms of this simple distinction the joint patterns within the Mbirao Group are of mixed type, i.e. diagonal, but in some cases with a pronounced 'single peak'.

In some areas the joint maxima parallel S_2 surfaces, or correspond with the theoretical axes of tension associated with S_2. For example east-south-east maxima occur in areas 1, 3 and 6 in Figure 19. Possibly these were formed as release tension-joints (De Sitter, 1956, p. 131), normal to the principal regional stress which was associated with the impression of S_2 in depth.

A swarm of close joints trending north-north-west is well developed in the lower Alivaghato Valley (sub-area 5B), broadly parallel to S_4 surfaces; these could similarly be interpreted as release tension-joints, but related to folding (F_4 phase) which affected the Lower and Middle Miocene sediments. Joint maxima trending north-north-east are a feature of central Guadalcanal, notably of sub-area 5A, which has a distinctly 'parallel' joint pattern, also of areas 1 and 4 and sub-area 5B. These joints are parallel to the axes of the anticlinal horsting, which affects the Pliocene sediments of north-central Guadalcanal, and therefore may be interpreted as Pliocene tension features.

Regular jointing is poorly developed in the Neogene sediments and volcanics, and is rarely observed in the Plio-Pleistocene; Thompson (1960) remarked on the absence of jointing in the Gallego Lavas. Only 90 joint-plane measurements have been made for areas 9 and 10 of north-central and north-eastern Guadalcanal. The pattern in area 10 is distinctly of 'parallel' type, with a north-north-west maximum presumably related to basement faulting during a relaxation phase subsequent to the F_4 folding. Local diagonal tension joints in the Mongga River, trending east, have been filled with siltstone from overlying strata, indicating initiation penecontemporaneous with the deposition of the lower part of the Mbokokimbo Formation, i.e. possibly Upper Miocene or Lower Pliocene.

In the Itina River area of south-west Guadalcanal, joints trending north-east in the Kavo Greywacke Beds may be regarded as cross-joints with respect to the system of north-west-trending folds.

The pattern in north-central Guadalcanal, area 9, contrasts with that of area 10: joints trending north-north-east are prominent, featuring particularly in the Gold Ridge Volcanics and Mbalisuna Gabbro. It is interesting to note that the joint maxima in these two sedimentary sub-areas parallel the local geophysical discontinuities (Figure 24), indicating structural controls in the underlying basement.

There is an impressive complex diagonal joint pattern in the Koloula Diorite (Figure 19, sub-area 8B). In view of the generally poor development of jointing in the Plio-Pleistocene sediments and volcanics, the jointing in Koloula must be related to local internal stresses occasioned by emplacement of the diorite, rather than to a regional stress pattern. Shearing due to horizontal stresses has occurred along zones of closely spaced joints trending east-south-east, for example in the Ghecha River, giving rise to zones of brecciated dioritic 'grush' which appear to have been formed while the intrusion was still hot and plastic, although sufficiently rigid

to be ruptured. The main outcrop of the Koloula stock is elliptical, its axis trending north-east; assuming that the diorite was emplaced into a north-east-trending tension feature, the jointing pattern could be analysed in terms of maximum horizontal confinement of the ascending pluton in a north-west to south-east direction, and minimum horizontal confinement at right angles, i.e. along the axis of tension, rather than vertically (compare Carey's interpretation, Figure 12A). This north-east-trending axis of tension parallels the fault swarms of Savulei which are associated with Pleistocene–Recent vulcanicity.

The joint and mineralised-vein patterns for the Gold Ridge Volcanics and the Koloula Diorite correspond in some detail, showing that fracture analysis may have an important bearing on mineral exploration in the Tertiary igneous domains of the Solomons.

Coleman (1957) considered the jointing in western Guadalcanal to be associated with the many minor faults. Pudsey-Dawson and Thompson (1958) concurred with this view and recognised two systems of joints, the main system trending north-west and north-east, and a subsidiary system trending north and east, the latter being associated particularly with the Ghausava Ultrabasics.

Certainly elsewhere in Guadalcanal joints are in many instances demonstrably associated with faulting; however, in view of the structural complexity of the Mbirao Group in particular, interpretations of joint patterns in terms of changing regional stress systems must be regarded as essentially tentative.

Discordant minor intrusives

The trends of 76 minor intrusives are represented on a rose diagram (Figure 20). The majority of the dykes are nearly vertical, although some may be inclined at angles as low as 50° to the horizontal. They vary in thickness from about 2 to 120 m.

Figure 20 Rose-diagrams showing the strike of discordant minor intrusives in Guadalcanal

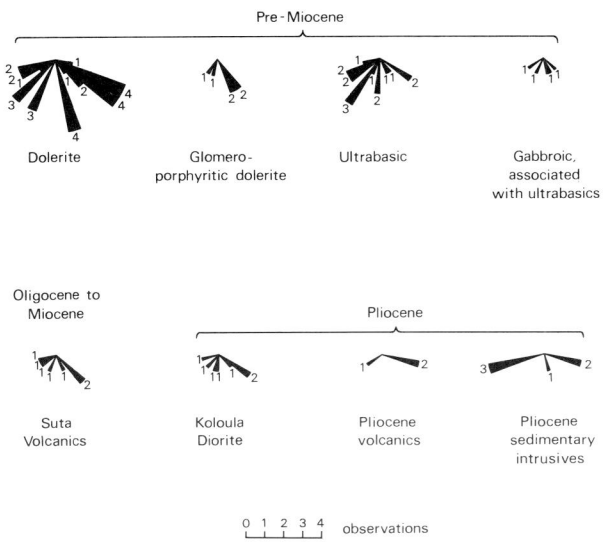

Table 9 The sequence of tectonic events in Guadalcanal

Epoch	Metamorphic s-surfaces	Emplacement of ultrabasic rocks and uplift	Folding
Pleistocene–Recent			Minor drag folds adjacent to Veuru Moli Fault
Pliocene	S_4: NNW–SSE in south-central and eastern part of Mbirao Metabasics belt	Uplift of over 2500 m	Open modal folds predominantly synclinal, affecting S_1, S_2 and L.–M. Miocene sediments. Secondary folds trending NNE–SSW in central Guadalcanal
Upper Miocene	S_3: NW–SE in Malango area south-central and eastern part of Mbirao Metabasics belt	Emplacement of Ghausava Ultrabasics; re-mobilisation of earlier emplacements	
Oligocene	S_2: WNW–ESE in central and eastern part of Mbirao Metabasics belt	Remobilisation	F_2: minor folds affecting S_1 (Mbirao Metabasics) and S_0 (Tetekanji Limestones)
Eocene	S_1: NE–SW along northern margin of Mbirao Metabasics belt	Emplacement of Suta and Marau Ultrabasics	Mbirao Geanticline. F_1: minor intrafolial folds
Pre-Eocene	Initiation of Mbirao Geanticlinal Welt and extrusion of basaltic lavas		

Within the Mbirao Group, in the headwaters of the Talise River, two parallel glomeroporphyritic dolerite dykes can be traced for a distance of 1 km along the strike; elsewhere single intrusives have not been traced over distances in excess of a few metres. Even allowing for exposure failure, it is apparent that the dykes are not persistent features, neither do they form close swarms of parallel trend in any area, although the trend frequency diagram shows two distinct maxima in the north-west and north-east quadrants.

Sporadic tension features such as dykes could theoretically be orientated in many different directions in relation to the maximum regional stress, particularly if second-order features are considered in relation to strike-slip faulting; accordingly it is not practicable to analyse all the dyke trends consistently in terms of relief of regional stress (Anderson, 1951, p. 22).

In the Mbirao Group it is presumed that most of the minor intrusives were emplaced in a pre-tectonic phase; for example, evidence indicates that the basalt dykes in the Guadalcanal Gabbro of the Valasi Block were intruded while the gabbro was still hot, although already rigid. Eight dykes have been mapped in this area, five of which trend north-east and dip at steep angles to the north-west; it is suggested that they were intruded in an initial phase of north-west to south-east regional stress, filling tension fractures in the 'suprastructure'.

Dolerite bodies in the Guadalcanal Gabbro and patches of metagabbro within the Mbirao Metabasics have been mapped as trending broadly north-east, for example in the Vaghanambo River above Vale Muse. It is not clear from the field relationships whether these were originally intrusive into the metavolcanics or represent tectonic slices of coarser basic rock thrust in from depth.

By way of contrast, five of the six glomeroporphyritic dolerite dykes mapped in the south-west part of the Mbirao Block trend north-north-west; these intrusives have been sheared during the third phase of tectonism, the glomerocrysts having been deformed into ellipsoidal shapes whose AB planes parallel S_4 surfaces in the surrounding metavolcanics. These dykes could also have been initiated as tension fractures in response to a north-westerly regional stress.

Intrusives associated with the ultrabasic masses include:
1 Minor ultrabasic dykes generally parallel to the trend of the main body, and intrusive into gabbro on the 'hanging wall side.'
2 Gabbroic dykes parallel to the trend of the ultrabasic boundary, for example in the upper Kuma Valley.
3 Dykes of metagabbro or pyroxenite, trending obliquely to the strike of the serpentinite boundaries
There is again a marked tendency for these dykes to trend north-west or north-east.

Faulting	Jointing	Discordant minor intrusives	Mineralisation
Oblique-slip step-faulting in Weather Coast and Veuru Moli zones; Savulei Fault Swarm, tension faults associated with Koloula Diorite	Diagonal jointing in Koloula Diorite		Copper sulphides in Koloula Diorite; veins parallel to joint pattern
Faults trending NNE and NNW in central Guadalcanal; reverse faults bounding fan-shaped horsts (e.g. Malango Axis).	Joints trending NNE in N-central Guadalcanal and release tension-joints in S-central and NE Guadalcanal	Dykes trend NE and NNW, parallel to major fault trends	Gold-quartz-pyrite mineralisation in Gold Ridge Volcanics and Sutakiki Valley
Initiation of Malango Fault (as first-order sinistral wrench?); formation of numerous minor faults; existing incorporated into changing strain pattern; initiation of Chimiu Fault Zone		Shearing of glomeroporphyritic dolerites in the Mbirao Group	Quartz and magnetite lodes trending NW in central Guadalcanal; mineralised zones associated with Poha Diorite trend NNW
	Release tension-joints parallel to S_2 in the central part of Mbirao Block		Pyrite, minor chalcopyrite and hematite veins in central Block
Initiation of Valasi Fault Zone (as first-order dextral wrench?)		Early intrusion of basic dykes in the Mbirao Group	Sulphide segregations on S_1 surfaces in the Mbirao Metabasics; quartz-chlorite-epidote veins as secondary tension in Mbirao Volcanics

Later andesite dykes, notably those associated with the Gold Ridge Volcanics and the Koloula Diorite, parallel major fault trends, and do not appear to be closely related to the local joint patterns. In the Koloula Diorite and surrounding Suta Volcanics, the majority of dykes trend north-west, i.e. parallel to the Koloula and Suta Faults.

Thus the pattern of minor intrusives associated with the Pliocene volcanics and the Koloula Diorite appears to resemble that of the older basement, in spite of the inferred changes in direction of regional stress since the early Tertiary. Possibly major tension features at depth, which had been initiated during the earlier (Mbirao) phases of tectonism, tended to be utilised as magma channels during the later phases, so that the fracture pattern of the 'infrastructure' was reproduced to some extent in the Plio-Pleistocene.

An interim tectonic synthesis

A synthesis of the structural data in terms of tectonic 'domains' of relatively homogeneous structural style is presented in Figure 8.

Essentially the pre-Miocene basement is divisible into two structural units: a schistose infrastructure, with ultrabasic emplacements; and a non-schistose suprastructure, e.g. the Valasi and Weather Coast Blocks.

The volcanic and sedimentary blanket includes the gently folded Oligocene–Miocene sequence, which rests unconformably on the pre-Miocene basement; and the unfolded Pliocene–Recent sequence, locally intensely faulted and pierced by Neogene intrusives.

Table 9 gives an indication of the sequence of structural events: four major tectonic phases are interpreted as shown on the stratigraphic chart (Figure 4).

The key to this synthesis is the sequence of events as indicated by the analysis of metamorphic s-surfaces in the pre-Miocene basement and the attitude of bedding planes in the overlying sediments. The fracture patterns may be related to this scheme in a general way, for example in the case of faulting, by reference to the Moody and Hill scheme; it cannot be claimed, however, that all mesoscopic features can be integrated unequivocally into a grand synthesis.

The timing must be regarded as tentative; in a given area structural features assigned to a particular phase may overlap into adjacent phases in time. The sequence may be summarised as follows:

Phase 1 (Eocene) The formation of the Mbirao Geanticline in response to north-west to south-east stresses, the emplacement of ultrabasics in the axial zone, and the impression of the earliest schistosity.

| Phase 2 (Oligocene) | Modification and 'remobilisation' of the Mbirao Geanticline in response to north-north-east to south-south-west stresses. |
| Phases 3 and 4 (Upper Miocene to Pliocene) | Further modification of the Mbirao Geanticline in response to north-east to south-west stresses; folding of Lower Miocene sediments. |

Vertical uplift and faulting have continued to Recent times, although the geomorphology of the present-day coastline of Guadalcanal suggests an interim period of relative stability.

FOUR

Economic geology of Guadalcanal

Since 1950 prospecting efforts in Guadalcanal have con-concentrated mainly on the search for copper and gold. Minor shows of nickel, iron ore, manganese, lead, zinc and bauxite have not as yet encouraged detailed investigation by private companies.

Deposits of economic interest, as indicated on the map (Figure 22) may be classed as follows:

1 Deposits associated with the pre-Miocene Mbirao Group:
copper sulphide deposits of marine volcanic association are here tentatively separated into two categories:

(a) Low-grade disseminations and minor segregations in the oceanic tholeiites of the Weather Coast Block and their metamorphosed equivalents (for example the Talise–Chimiu area and the occurrences in the upper Tina and Ghausava rivers).

(b) Lenses of massive exhalative volcanogenic sulphide in the Valasi Block, considered by O'Sullivan and others (1975) to be associated with a basaltic suite of ophiolitic affinity, in contrast to the basalts of the Weather Coast Block and the Mbirao Metabasics, which are of primary oceanic association. The copper sulphide occurrences of the Ghari area in south-west Guadalcanal might also be considered under this category. Non-cupriferous deposits include:

(c) Manganese oxides as exhalative volcanogenic products associated with the Tetekanji Limestones (for example Na Vuvuti, mentioned by Grover, 1965).

(d) Gold deposits of the Sutakiki Valley.

(e) Residual concentrations of nickel in lateritic soil on the ultramafics (for example Marau, Suta and Ghausava).

2 Deposits associated with the post-Eocene felsic igneous association:

(a) Diorite complexes of 'porphyry copper' type (for example the Koloula Diorite, the Poha Diorite, and possibly also the Mbetilonga occurrence).

(b) Low-grade disseminated sulphide deposits associated with andesites and high-level porphyrites of the Gallego Volcanics (for example Hidden Valley and other minor occurrences in western Guadalcanal).

(c) Low-grade exhalative and epithermal gold deposits associated with andesitic pyroclastics (the Gold Ridge area of central Guadalcanal).

(d) Alluvial gold deposits in the gravels of the lower Matepono River, derived from the Gold Ridge deposits.

(e) Possibilities of gold in other areas.

3 Hydrocarbon prospects, for example for petroleum and lignite.

4 Residual concentrations of (a) bauxite on calcareous formations and (b) iron, as, for example, at Cape Henslow.

5 Beach-sand prospects, for magnetite and ilmenite.

6 Cementstone prospects, and occurrences of volcanic ash with 'pozzolanic' properties.

A detailed discussion of the geological relationships of these prospects is beyond the scope of this memoir. Only the general associations are described below, together with citations of references relevant to the history of prospecting.

On consideration of detailed structural evidence, different phases of mineralisation have been tentatively integrated into the tectonic history, as shown in Table 9.

Rose diagrams for trends of mineralised veins within the Mbirao Group are given in Figure 21. Only persistent vertical or quasi-vertical structures have been included.

Deposits associated with the Mbirao Group

Copper in oceanic tholeiites of the Mbirao Volcanics

Fine-grained sulphide, predominantly pyrite, is commonly developed on surfaces of schistosity within an area of about 120 km^2 in south-central Guadalcanal; it is particularly evident in the headwaters of the Talise, Alivaghato and Kolom-

bolavu rivers. Analyses of 366 samples of river sand collected by the writer gave a background of 150 ppm for hot-extractable copper over the Mbirao Metabasics outcrop; locally values for total copper in soils reached 650 ppm. The background value is somewhat higher than the arithmetic mean of 119–123 ppm Cu for basalts quoted by Prinz (1967). Finely disseminated sulphide is also associated with metabasics in the Tina headwaters (Grover, 1958c). Only thin veinlets of chalcopyrite and magnetite have been located *in situ*.

Within the Mbirao Group there is a strong tendency for the strike of mineralised fractures to parallel the predominant local structural 'grain' as defined by swarms of faults and metamorphic s-surfaces.

In the Weather Coast Block (Figure 21, areas 1 and 2) veins of quartz, chlorite and epidote occur as *en échelon* zones of tension gashes, or as individual tension fractures which may be traced for as much as 50 m along the strike; the axis of tension in a majority of instances trends east–

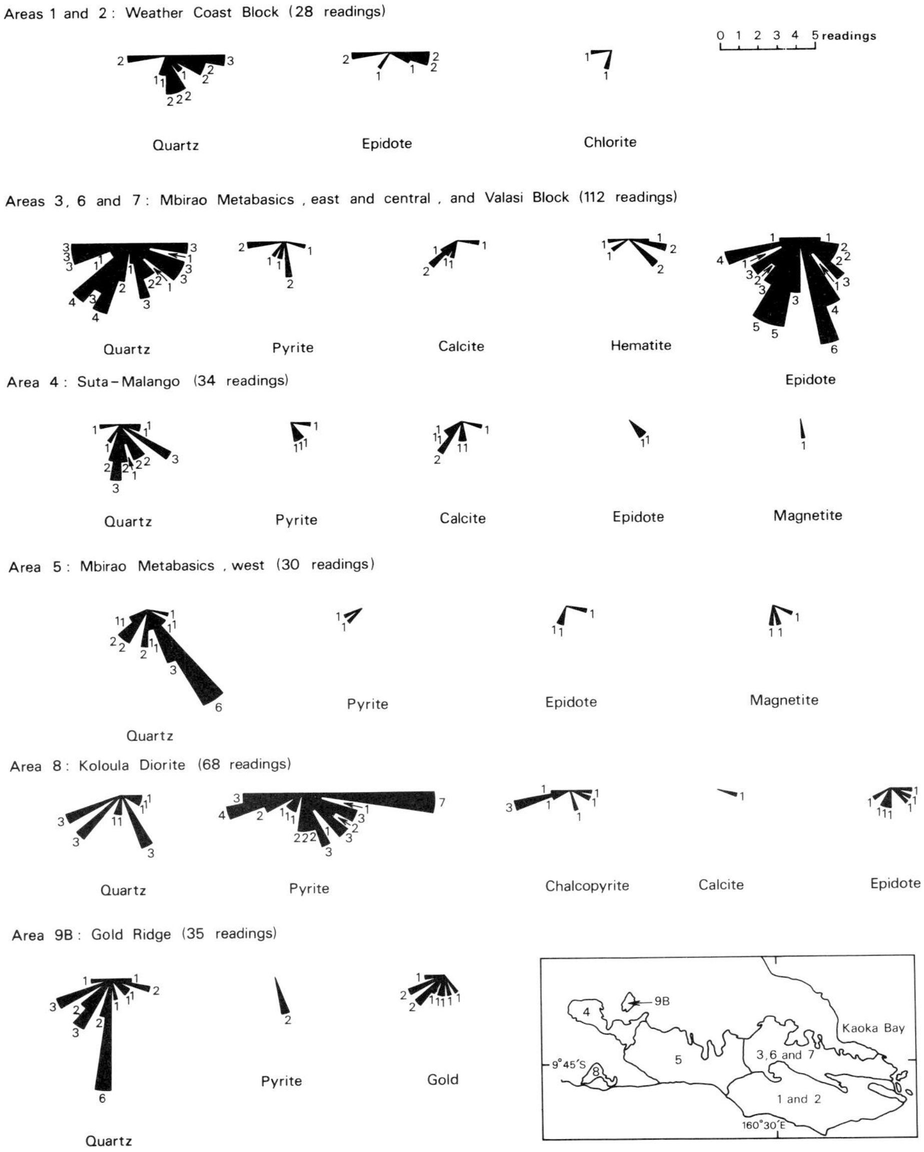

Areas 1 and 2 : Weather Coast Block (28 readings)

Quartz Epidote Chlorite

Areas 3, 6 and 7 : Mbirao Metabasics , east and central , and Valasi Block (112 readings)

Quartz Pyrite Calcite Hematite Epidote

Area 4 : Suta–Malango (34 readings)

Quartz Pyrite Calcite Epidote Magnetite

Area 5 : Mbirao Metabasics , west (30 readings)

Quartz Pyrite Epidote Magnetite

Area 8 : Koloula Diorite (68 readings)

Quartz Pyrite Chalcopyrite Calcite Epidote

Area 9B : Gold Ridge (35 readings)

Quartz Pyrite Gold

0 1 2 3 4 5 readings

Figure 21 Rose-diagrams showing the strike of mineralised veins in eastern and central Guadalcanal

west, although there is a subsidiary maximum trending north-north-east. Both maxima could be regarded as second-order tension effects with respect to the first phase of tectonism in the Mbirao Group, related to S_1. During this phase quartz and epidote would have been partially remobilised from interstitial aggregates within the pillow lavas.

In the Mbirao Metabasics the pattern of mineralisation is more complex; quartz-epidote-calcite venation is predominant in central and east Mbirao (areas 3, 6 and 7). In the headwaters of the Kolohaisava River narrow veins of pyrite and hematite are associated with the northern marginal zone of the Weather Coast Block, and parallel the line of the Tetekanji Fault.

In west Mbirao (area 5) there is a somewhat higher background of copper in the river sand but most of the sulphide is very finely disseminated in the metavolcanics. In the Kolombolavu and Talise headwaters S_1-surfaces are typically coated with very fine scales of crystalline sulphide,

predominantly pyrite. There is, however, a tendency for quartz and magnetite lodes to trend north-west, that is parallel to S_3-surfaces and presumably related to the same phase of tectonism; occasionally traces of copper sulphide are associated with the magnetite lodes.

On the basis of anomalies in beach sands at river mouths, Pisarski (1968), as part of the UNDP Aerogeophysical Survey Project, delineated a 'Copper Province' on the Weather Coast, which embraced the entire drainage systems of the rivers between the Ghalighecha and the Manauvo. This area virtually coincides with the Weather Coast Block plus the Koloula Diorite and associated Suta Volcanics.

Parts of the Weather Coast Block have been prospected successively by CRA Exploration Pty Ltd (Talise–Chimiu, Mackenzie, 1968a), Utah Development Company (Avuavu area; Jarvis, 1970, and Gardner, 1972b) and again by CRA (O'Sullivan and others, 1975). O'Sullivan preferred to regard the Weather Coast Block and the Mbirao Metabasics as representing a suite of oceanic flood basalt affinity,

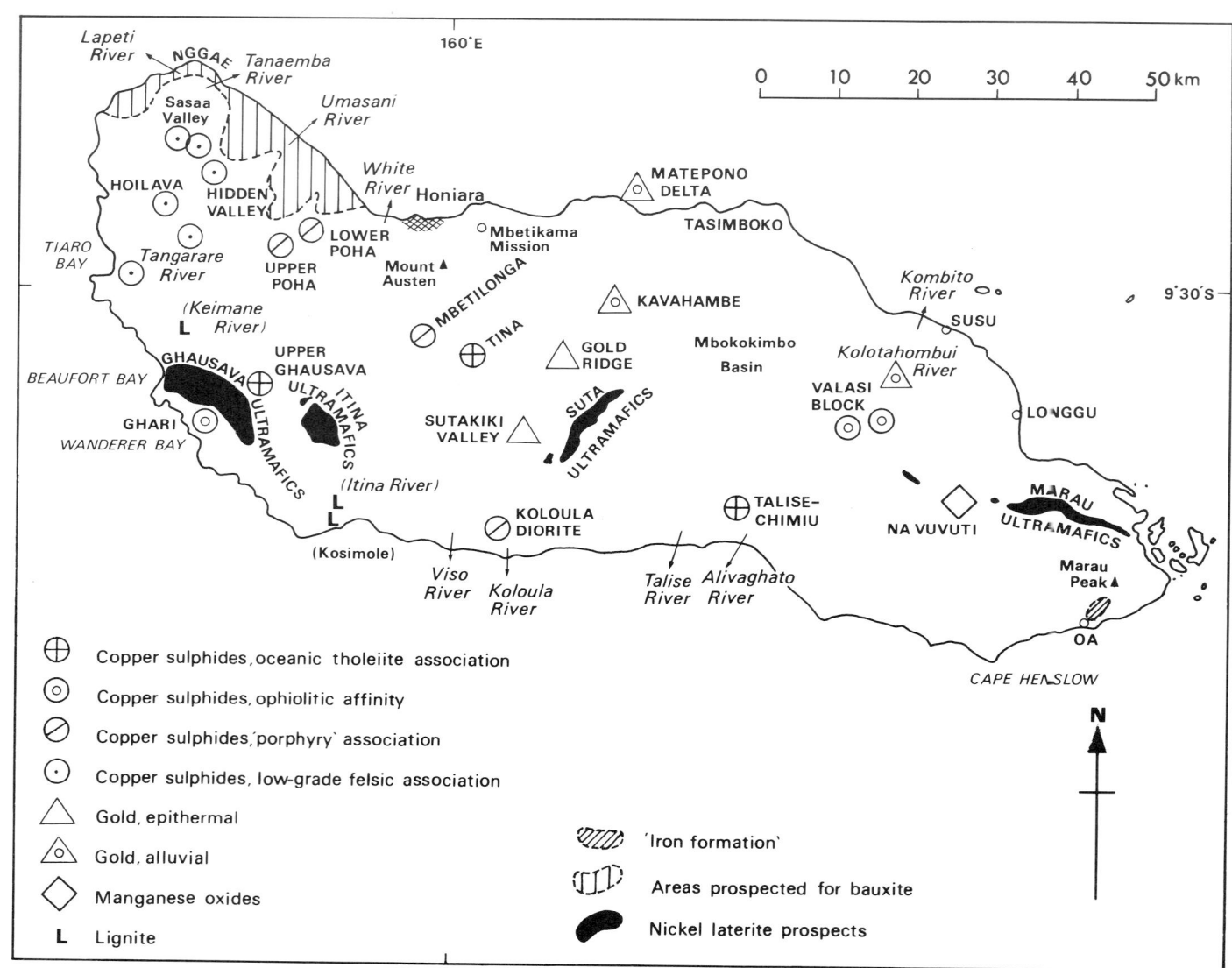

Figure 22 Map of prospects for economic minerals in Guadalcanal

the Mbirao Mafic Igneous Suite, as opposed to the Valasi Block, which might be considered to be essentially of ophiolitic association (the Valasi Mafic Igneous Suite). CRA's results show a range of copper in the rock and float for the Mbirao Suite of 10–1400 ppm. For stream sediments a median value of 170 ppm Cu was quoted for an approximate background value, with values of above 500 ppm being considered anomalous. Significant concentrations of copper were judged to be unlikely in the Mbirao Igneous Suite.

Further west, Grover (1958c) described a pyrite body, assaying only up to 0.8 per cent Cu, in sheared metabasics at the Mbicho/Tasi confluence in the Tina headwaters. Grover (1965) also referred to R. B. Thompson's descriptions of sulphide in sheared Mbirao Volcanics in the Ghausava headwaters. The latter area was included in a geochemical prospecting programme by Carpentaria Exploration Co. Pty Ltd (Coles, 1971); no significant copper values were found in the stream sediments.

Turner recorded extensive quartz veining and pyritisation in the Chembea tributary of the Ghausava (Turner and Hackman, *in preparation*).

Massive sulphide deposits in the Valasi Block

O'Sullivan and others (1975) described massive exhalative volcanogenic sulphides, commonly associated with brecciated basaltic flows of the Valasi Block (Valasi Mafic Igneous Suite). Significant sulphide was only found in the Valasi River (notably the Mbombo tributary) and the Rere: rock chips assayed up to 2.9 per cent Cu, *in situ* lenses of variable massive sulphide being recorded up to 1.5 m in thickness. Chalcocite float found in the Kombito River assayed up to 8.4 per cent Cu and 280 ppm Zn. However, the mineralised area was considered to be too limited to warrant further prospecting.

In general, the Valasi Suite mineralisation shows higher values of Ni, Co, Cr and Mn than the Mbirao Suite; however, more whole-rock and trace-element analyses are required before separation of the two mafic igneous suites can be justified on petrochemical grounds.

The Valasi Block is transected by swarms of north-east trending epidote veins, tentatively correlated with the first (S_1) phase of tectonism.

Minor pyrite/chalcopyrite lodes trending north–south in the Vaghanambo River below Vale Muse (on the margins of the Valasi Block) have been sheared along planes parallel to S_4. As these veins occur on either side of the Tetekanji and Valasi faults, in parallel sets, it may be inferred that they were emplaced during a hydrothermal phase which post-dated the metamorphism of the Mbirao Metabasics (S_1) and the activation of major east-south-east faults (S_2); where transected by S_3–S_4 surfaces, these veins have been sheared—furthermore, as mineralisation has not been observed in the Lower Miocene sediments, which were folded during phase 4, the mineralisation in central Mbirao can be assigned tentatively to phases 2 to 3, probably late 2 (Table 9).

Craig (1968) described a multiphase basaltic/doleritic complex outcropping between the Ghausava Ultramafics and the Mbirao Volcanics in the hinterland of Wanderer Bay in the Ghari district of western Guadalcanal.

Sulphide float, including pyrite, bornite, chalcopyrite, chalcocite and malachite, has been traced to impersistent and sporadic lenses in the mafic complex, often associated with narrow veins of milky quartz.

Newman (1968) studied part of the prospect in the Ghunukeo (Vunikeo) tributary of the Charikumbau River; Amoco Minerals Solomons Ltd (Barrus, 1975) prospected the area in 1974 but considered it unworthy of further attention. The occurrence appears to show some similarities of association with the Valasi Block mineralisation. Turner and Hackman (*in preparation*) gave more details of prospecting in this area.

Manganese

In the Na Vuvuti area of eastern Guadalcanal sporadic pockets of grey powdery manganese wad occur in the Tetekanji Limestones (Grover, 1965). Both the limestone and adjacent brecciated basalt have been stained by finely divided iron and manganese oxides. However, preliminary analyses have not yielded concentrations in excess of 2.5 per cent Mn, and the remoteness of the area has discouraged further investigations.

Gold in the Suta area

Grover (1955h) described the gold prospects of the Sutakiki tributary of the Sutakama. Between 1937 and 1940 a small syndicate attempted hydraulic sluicing of the river bed, but the work was abandoned owing to frequent floods and logistic difficulties. Detailed mapping of the area (Hackman, 1968c) demonstrated that the country rock was sheared and altered pre-Miocene Guadalcanal Gabbro. The gold appeared to be associated with sulphide in minor quartz reefs; arsenic determinations were made on stream sediments in the hope that it could be used as a pathfinder for gold. Several arsenic anomalies were indicated in the lower Sutakiki and in the Sutakama downstream from the Sutakiki confluence.

The venation pattern in the Suta–Malango area (Figure 21, area 4) shows two distinct maxima; quartz-pyrite-epidote veins trending north-west, and quartz-calcite veins trending north. Gold has been found in the Sutakiki Valley associated with quartz lodes, and it is interesting to note the correspondence with the gold-quartz-pyrite mineralisation pattern of the Gold Ridge Volcanics (sub-area 9B): the latter diagram incorporates some information derived from Grover (1958b).

Grover (1965) considered that the area would be worthy of further prospecting should road access from the north be improved as far as Gold Ridge.

Nickeliferous laterites

The ultramafic outcrops of the Marau and Ghausava areas were prospected by International Nickel Southern Exploration Ltd in 1962 and again in 1970 for both nickel and cobalt (data held on file in Ministry of Natural Resources, Honiara). The laterites are photogeologically distinctive, being vegetated with fern and light scrub.

On the Ghausava Ultramafics the highest value for Ni+Co obtained from laterite samples was 1.56 per cent, and the most promising area was the western half of the block. Prospecting in the Marau area had to be abandoned in 1970 due to opposition from the local inhabitants. The Suta Ultramafics are too deeply dissected and difficult of access to be of economic interest.

There are no indications of garnierite deposits, or of massive nickel sulphides associated with these Alpine-type serpentinised harzburgites.

Turner and Hackman (*in preparation*) present some of the hitherto unpublished assay results in tabular form.

Post-Eocene felsic igneous association

Prospecting in the younger Cainozoic volcanic province of western Guadalcanal has revealed the presence of low-grade sulphides in veins and disseminations associated with high-level dioritic intrusives. The 'porphyry copper' association is probably best exemplified by the Koloula Igneous Complex.

Koloula Diorite

Grover (1965 p. 60), referring to Pudsey-Dawson's field notes, reported chalcopyrite in quartz veins, and, in particular, gold from the Kosikosi tributary of the Koloula River. Later work has not confirmed the presence of gold. Although the area was too rugged to be flown by the UNDP aerogeophysical survey, Winkler (1968a) outlined the results of drilling and other geological work, which led to the delineation of a geochemical anomaly exceeding the threshold value of 1350 ppm hot-extractable Cu over an area of 9 ha straddling the Ghecha valley.

In 1969 Utah Development Co. commenced prospecting (Warin, 1969). The programmes, involving geological surveys, geophysical ground exploration (induced polarisation), petrographic investigations, geochemical sampling and drilling, continued until 1974. It was found that dispersion trains for hot-extractable Cu were very short, being traceable for no more than 500 m (the alkalinity of the surface waters was thought to be responsible).

Prospecting was concentrated in two areas: Chikora, in the Ghecha Valley, and Mbina in the upper Koloula. However, 1830 m of diamond drilling in 1974, in four separate holes, encountered no economic grades. Further explorations could not then be financed.

The writer found in his preliminary survey that vein mineralisation and jointing patterns appeared to be closely related. Sulphide veins in the Koloula Diorite show distinct frequency maxima at azimuths of N070° and N090°, the former are predominantly tension fractures, whereas shear effects have been observed in association with the latter.

Mineralisation in the altered quartz-diorite zone has been described by Youles in Winkler (1968a); pyrite and chalcopyrite are sparsely disseminated, but discrete lodes of pyrite, with chalcopyrite, bornite and chalcocite are conspicuous in the Ghecha Valley and parts of the upper Koloula. Frequently quartz forms a selvage to the sulphide lodes, and malachite a secondary coating.

Chivas (1975) has examined the petrography of the complex in considerable detail. He has distinguished two major suites, a Mafic and a Felsic Suite, comprising eleven intrusive phases in the concentrically zoned pluton: the composition ranges from leucogabbro to granodiorite. Two distinct hydrothermal episodes have given rise to copper mineralisation. Further age determinations on the granodiorites confirm ages of 1–2 million years.

Chivas and Wilkins (*in press*) suggest that the abundance of fluid inclusions may be directly related to the degree of mineralisation, and that the western and northern parts of the complex, where microfracturing is more intense, may be better prospects for copper than the southern half.

Poha Diorite

The Poha Diorite, only 15 km west of the capital, Honiara, has been prospected intermittently since the early 1950's. However, most of the prospecting has been concentrated on the more accessible areas of interest in the lower and middle Poha valleys. Hackman (1977) describes the petrography and history of prospecting in some detail.

Pudsey-Dawson and Thompson (1958) noted floaters of cupriferous sulphides and magnetite in the streams between the Mavo and the Poha.

Grover (1958e and 1965) drew attention to flat-lying tabular bodies of magnetite in the Poha–Sura headwaters.

Indications of low-grade sulphide mineralisation are scattered in widely-separated areas within the Poha Diorite Complex. In 1963 E. Zohar (written communication) delineated copper anomalies in the soil and stream sediment in the middle and lower Poha Valley. He considered the mineralisation to be unevenly distributed as disseminations in veinlets and fissures throughout the complex.

Subsequently the area has been prospected in variable detail by Hanna-Homestake Mining Co., CRA Exploration Pty Ltd, Carpentaria Exploration Co. Pty Ltd, and Utah Development Co. A large body of geochemical exploration data has accumulated, the value of which is difficult to assess owing to the fact that different programmes have employed different methods for the preparation of samples and determination of copper. Perusal of a final report on prospecting by Amoco (Barrus and Weiland, 1975), specifically on exploration in the middle Poha Valley, shows that there is some correlation between copper anomalies in the soil and the intensity of fracturing in the diorite. Exploration was discontinued, although further prospecting was recommended for the Siu Valley area, on the basis of anomalies associated with a 'pebble dyke'.

From the upper Poha–Sura area, still not prospected in detail, Hackman (1977) reports assays of zinc up to 9 per cent in floaters of sphaleritic and cupriferous material.

As the UNDP aerogeophysical survey had delineated a radiometric anomaly over the Poha Diorite in the Mavo–Mbonehe headwaters and in the adjacent Umasani Valley, the area was reconnoitred in 1967. The mineral torbernite had been recorded in a floater from the Mavo River (Pudsey-Dawson and Thompson, 1958), and the possibility suggested itself that copper might occur in association with uranium in that area. Tests for uranium n the water of the upper Mavo and adjacent Umasani tributaries proved negative.

Zohar (oral communication, 1963) recognised the association of zones of disseminated copper ore with faults trending north-north-east. As the Poha Diorite and surrounding volcanics have not been subjected to the earlier phases of the tectonic history, the mineralisation in this part of Guadalcanal could conceivably be assigned to phase 3 (Table 9).

A close comparison may be made between the Poha Diorite and the Koloula Diorite (Mason, 1975). Although the core of the Koloula Diorite is more acidic than the Poha tonalite, the range of rock type is similar, suggesting hybridisation.

Also the relationships between the Koloula Diorite and the propylitised Suta Volcanics parallel those between the Poha Diorite and the Umasani Volcanics, although the Poha Complex completed its history of intrusion before the Miocene, the Koloula complex in the Pleistocene. In both cases the primary foliation and orientation of melanocratic bodies trend north-east, that is at right angles to the predominant structural trends of the Solomons Arc.

Mbetilonga

The Mbetilonga Basin is 12 km south of Honiara. Grover (1955a) first mapped the area in 1952 and reported veins of 'ferruginous, cupriferous, siliceous material' in a metamorphosed pyroclastic formation, associated with quartz-diorite. Earlier, in the 1930's, the Mbusasangatu tributary of the Mbetilonga River had been prospected for alluvial gold by Anton Olsen.

Hill (1960) mapped the Mbetilonga Basin in some detail: he described the 'basement' rocks, hosts to the mineralisation, as sheared agglomerates with intrusive diorites, and also delineated sporadic patches of mineralised rock and the locations of boulders of siliceous 'lodestuff', which proved to be predominantly cryptocrystalline quartz.

Newman and Youles (1968, written communication) carried out a geochemical exploration programme involving ridge-crest soil sampling and a limited stream-sediment survey. They suggested that soil anomalies might be related to sporadic lenses of chalcopyrite associated with pyrite, and recommended drilling in the Lotu area to the east of Mbetilonga.

In 1964–65 CRA Exploration Pty Ltd conducted a geochemical drainage reconnaissance in the Mbetilonga Basin and adjacent areas of the Lungga Valley.

Mackenzie (1968b) recorded values of up to 1350 mg/kg Cu in stream sediments and up to 4650 mg/kg in soil samples. Some soil-profile analysis was carried out. He suggested that higher copper values in the soil tended to occur near the projected contact between the 'basement' and the overlying Lower Miocene limestone, and were possibly explained by concentration during a period of pre-Miocene weathering. The areal extent of soil values above 2500 mg/kg was too limited to justify further work.

In 1972–73 Utah Development Company organised a more extensive exploration programme (Gardner, 1972a and 1973). Soil sampling in the Mbetilonga Basin and adjacent Mbetisahata Valley led to the demarcation of ground anomalies which were tested by induced polarisation and apparent resistivity techniques. A drilling programme followed, but further exploration was not financed.

Grover (1955a) described the mineralogy as predominantly pyrite, quartz and chalcopyrite, with minor development of bornite, tenorite and malachite. Very minor traces of silver were assayed. Veins of barytes and magnetite were recorded, also talc and traces of very finely divided alluvial gold. Grover (1965) also referred to veins of alunite and diaspore, traces of sphalerite, hematite, goethite and apatite. Gardner (1973) added chalcocite and covellite to the list of supergene copper minerals, also remarked on the abundance of limonite and the fact that gypsum was often associated with sulphide; the presence of talc was not confirmed.

Fresh rock exposures are rare in the Mbetilonga area—much is obscured by the extensive silicification and propylitisation. Gardner (1973) concluded that the host to mineralisation was a pyroclastic unit which was underlain by the 'basement' and overlain by the Mbetilonga Limestone. However, Hughes (*in preparation*) in the course of detailed remapping of the Lungga–Mbetikama area has distinguished between different facies of diorite, intrusive into basaltic andesites. It has emerged that on a regional basis the suite is broadly comparable with the Poha Diorite and the ambient Umasani (Suta) Volcanics which it intrudes. Relationships have been obscured by alteration and subsequent brecciation. It appears that the unit presumed to be pyroclastic is in fact derived from an andesitic–dioritic complex; the aspect of the association has been obscured by tectonic brecciation and shearing which post-dated the mineralisation.

Low-grade disseminated sulphides associated with the Gallego Volcanics

Certain minor copper prospects in north-west Guadalcanal have been investigated intermittently. As they are all located in the Tiaro Bay hinterland, they are described in some detail in the bulletin accompanying the geological map of that particular area (Hackman, *in preparation*).

The occurrence that has received most attention is that of Hidden Valley, in the headwaters of the Umasani River on the southern flanks of Mount Gallego. The principal mineralised area comprises a window of about 0.9 ha of propylitised porphyritic microdiorite or hornblende-andesite, an up-faulted inlier between the Gallego Lavas and the Tiaro Tuff-breccias with associated Pleistocene reef limestones.

Most of the exposed prospect is a white siliceous gouge, with sporadic pods of copper sulphide enrichment. Grover (1965) and Thompson (1965) described the geology of Hidden Valley, the mineralisation and a magnetic survey. Drilling was undertaken by Southern Mining and Development Company (Grover, 1968c) and the area has since been investigated by Carpentaria Exploration Co. Pty Ltd and Amoco Minerals Solomons Ltd.

The principal sulphide in the mineralised pods is pyrite, and chalcopyrite is subordinate. Some sphalerite is present, also hematite, limonite and secondary manganese oxide. Chalcocite/covellite enrichment is quite marked in surface breccia zones, but has been shown to be insufficiently thick to constitute an ore-body.

The mineralisation appears to be associated with high-level 'sub-volcanic' bodies of microdiorite within the Gallego Volcanics. It is possible that some massive sulphide pods represent slivers of Pleistocene reef limestones that have been preferentially mineralised.

Boulders of silicified 'lodestuff' similar to that found in the Mbetilonga area, were recorded by Pudsey-Dawson in the headwaters of the Hoilava and Tangarare Rivers (Grover, 1965). Geochemical exploration follow-up has not revealed associated base-metal anomalies in the stream sediments from the source area, although the zinc values are moderately high (300–840 mg/kg) in the middle Hoilava valley.

Some pyritisation, associated with values of up to 370 mg/kg copper in stream sediments, has also been noted in the immediate hinterland of Tiaro Bay. Possibly in that area mineralisation is associated with pre-Miocene volcanics of 'Suta Volcanics' type.

Gold deposits

Gold Ridge

Gold is associated, as low-grade disseminations and in fractures, with Pliocene clastic volcanics of predominantly basaltic andesite composition, which form an integral part of the Toni Lithosome. The area concerned is approximately 9 km^2. Located 29 km south-west of Honiara, the prospect is 4 km from the Omboombo roadhead, whence it is accessible on foot.

Grover (1955d and 1958b) has written much on the history of Gold Ridge. The presence of gold was recorded from the alluvial sands of the Matepono River in 1568 by the expedition of the Spanish explorer Alvaro de Mendaña: its occurrence at Gold Ridge was first confirmed by the botanist Kajewski in 1936.

The Balasuna Syndicate commenced exploration in 1938, and by 1941 numerous pits and adits had been constructed. Work was discontinued in 1942 due to the Japanese invasion.

Connolly (1948) collected samples over much of the Gold Ridge area: his three highest assays ranged from 10 to 45 dwt/ton (15–69 mg/kg) Au. A positive correlation was indicated between sulphur and gold.

Jones (1949) suggested that gold should be concentrated in the pyroclastics adjacent to lava flows and dykes, although later work by Walshaw (1974) revealed no evidence for the existence of either lava flows or dykes.

In 1950 the first official examination was made (Grover, 1958b), and the area was visited again in 1955. The results of analysis of stream-sediment samples from the Chovohio River system (Hackman, 1968c) suggested the possibility that arsenic might be used as a 'pathfinder' for gold.

Walshaw (1974) undertook the most comprehensive official examination of the area: this involved an extensive programme (April 1971 to November 1972) of stream-sediment and soil sampling followed by pitting and drilling. Samples were analysed variously for Au, Ag, Cu, Zn and As. As a result of this work Walshaw considered that there might be potential for an open-pit operation. CRA Exploration Pty Ltd prospected the area in 1974–75 (Purvis, 1975): geological mapping was undertaken, also auger sampling and a diamond drilling programme (five holes were drilled, totalling 502.9 m).

The results of Walshaw's geochemical exploration confirmed that there was some degree of positive correlation between gold and arsenic. The gold is often visible as minute discrete particles, primarily in association with quartz and chlorite: pyrite, arsenopyrite and manganese dioxide also occur. Gold-bearing quartz veins, varying in thickness from 1 mm to 12 cm, range widely in attitude, and have not been traced individually for more than 10 m along the strike. No large lodes have been found.

Walshaw's mapping of the alteration revealed an inner sericitised zone, and outer zone of chloritisation, and several zones of silicification: usually the matrix of the rock is preferentially silicified whilst the clasts are liable to be softened, weathering out in the streams to potholes. Quartz veins in the highly altered rocks may be associated with goethite, hematite, sericite, kaolinite, jarosite and traces of arsenic, copper and zinc.

Two major gold-bearing zones appear to coincide broadly with the quartz-sericite zone: gold is preferentially associated with patches of strong silicification.

Gold, silver and iron sulphide tend to be enriched in local patches which are associated with dominant fractures trending north-north-east.

The venation pattern in the Gold Ridge Volcanics as recorded by the writer is presented in Figure 21, sub-area 9B. In the Gold Ridge area most of the veins dip at fairly high angles, although some auriferous lodes trending north-east dip at 20°–40° to the south-east and might be associated with overthrusting towards the north-west, a local expression of asymmetrical horsting in the adjacent Tausoro area to the south-west.

The correspondence between the Gold Ridge and Suta patterns (Figure 21, area 4), suggests that the gold-quartz-pyrite mineralisation in both areas may have been synchronous, that is Pliocene in age (Figure 19).

To explain the genesis of the gold, Walshaw suggested a mechanism whereby Au, As, Sb and Hg were precipitated from very dilute solutions in a circulatory system maintained by thermal convection through the permeable 'infra-structure' of clastic volcanics.

Purvis (1975) considers that the gold mineralisation was essentially 'exhalative', associated with hydrothermal alteration penecontemporaneous with the tuffs and breccias in a near-surface environment, and hence stratigraphically controlled. The highest gold values are in fractures that resulted from epithermal mineralisation in the infrastructure. Over an area of 3 to 4 km^2 the general tenor of the ore is 0.25 to 2 mg/kg, locally rising to 5 mg/kg. The CRA report concluded that the inferred extent of the deposit is insufficient to make it an economic proposition.

Alluvial gold

Since the 1930's small-scale prospectors have won gold intermittently from the alluvial gravels of the Chovohio River valley below Gold Ridge. Local landowners continue to make a living by panning.

Grover (1955g) described the history of these workings, with particular reference to the Kavahambe (Kovagombi) flats 6 km downstream from Gold Ridge itself. He interprets the Kavahambe gravels as having been deposited in constricted estuarine conditions during a period of aggradation due to interruption of the fluvial cycle by submergence. The maximum depth of the gravel at Kavahambe is about 20 m.

Wells (1965) examined the Kavahambe area for Bulolo Gold Dredging Co. Ltd (Clutha Development Limited) in 1951, using a light 4-inch hand drilling rig. He estimated that there were 10^7 m³ of alluvial material, some with moderately high gold content, but the bulk of it being barren gravel, sand and silt. As the better paystreaks were deeply buried under coarse gravel, and the distribution of values was uneven, the prospect was rejected as being of no further interest since the average grade (0.066 dwt/yd³ or 0.134 g/m³) and total quantity available is insufficient. Stanton (1965) visited the area in the same year, and agreed basically with Wells' assessment. He also considered that the gravels downstream in the Matepono Valley of the Guadalcanal Plains were likely to be of even lower grade.

In 1975 Turner and others (1977) conducted a short near-shore bottom-sediment sampling programme for gold, covering the narrow sedimentary prism off the mouth of the Matepono River, and a buried river-channel to seaward up to 7 km offshore. Only three out of the 172 samples taken yielded detectable gold—up to 87 μg/kg. Sampling employed both a piston corer and bottom surface dredging.

Grover (1958d) panned a few traces of gold from the Kolotahombui tributary of the Kombito River in north-east Guadalcanal. B. Du Faur had drilled the flood-plains behind Susu village in 1934, but did not find any significant gold values. The writer was unable to confirm the presence of gold in this area: the gold could have been derived from the conglomerates of the Pliocene Vatumbulu Beds.

A colour of gold was also detected in a beach-sand sample from Longgu in north-east Guadalcanal (Grover, 1960).

Possibilities of gold in other areas

The results of Pisarski's (1968) geochemical exploration survey of river-mouth beach sands indicated gold anomalies in the Lapeti River of north-west Guadalcanal and in the Viso and Koloula rivers of south-central Guadalcanal. The Viso and Koloula anomalies can be attributed to the Koloula Diorite complex, and the Lapeti anomaly to the Gallego Volcanics. Follow-up investigations in these areas have not led to any significant discoveries.

Fossil fuels

Petroleum

Grover (1955f) referred to palaeontological evidence and structural conditions favourable for the accumulation of petroleum in the Solomons and suggested that the inter-insular sedimentary basins might provide suitable reservoirs.

In Guadalcanal the greatest thickness of sediment is attained in the north-central part of the island, under the plains. Coleman and Day (1965), commissioned by Oil Search Ltd, carried out geological field work and gravimetric traverses with a view to assessing the petroleum potential. They suggested that the Lower Miocene limestones might provide suitable reservoirs; however, the structure of the area is not really favourable, in view of the subsurface basement high indicated by the Tetere Anomaly (Figure 24), which implies restriction of the volume of the sediment in the western plains. The Mbokokimbo Basin to the east might be considered a better proposition; however, an area of only about 360 km² is suitable for effective prospecting.

De Golyer and Macnaughton (1965) consider that the attractiveness of the prospect is diminished by the fact that no oil production has been proved, and by the absence of traps such as anticlinal structures. The second objection cannot be sustained, however, as the writer's observations on the pre-Pliocene sediments show that they are in fact gently folded.

Van Deventer (1971) included this part of Guadalcanal in a general reconnaissance of stratigraphy, structure and facies for Bataafse Internationale Petroleum Maatschappij. According to his interpretation of the Mongga, Mbokokimbo and Mberande river sections, Pliocene arenites in the Mbokokimbo Basin directly overlie the lower Miocene calcarenites.

This interpretation, dependent on age ranges inferred from planktonic foraminifera, conflicts in some respects with the ages inferred from benthonic foraminifera according to Coleman (1960b), as used in the stratigraphic succession adopted in this memoir.

Lignite

Grover (1955e) described seams of lignite up to one metre thick from the Lungga Beds in the Uraghai tributary of the Itina River. Grover (1965) referred to these occurrences again, also to partly pyritised 'Lignite measures' in the Ghausava River headwaters and the Keimane tributary of the Ghausava. Thompson (1968) suggests that these different deposits may be correlated stratigraphically: none of them constitutes an economically viable prospect.

Residual deposits

Bauxite

UNDP aerial radiometric data for the island of Rennell, 200 km south of Guadalcanal, revealed the presence of anomalies of large areal extent associated with red bauxitised clay soil on an uplifted atoll limestone complex (Winkler, 1968c). Radiometric laboratory tests by Overseas Geological Surveys showed a concentration of equivalent U_3O_8 in the soils of 0.01 to 0.06 per cent. G. R. Taylor (oral communication) has recently (1975) established that the anomaly may be attributed to radium.

The proved association on Rennell of radiometric anomalies with bauxite deposits of economic interest led to the examination of scintillometer anomalies elsewhere in the Solomons. In Guadalcanal soil sampling was undertaken in 1967–68 on the narrow belt of Honiara Reef Limestones between White River and the Umasani, also on the Mbonehe Limestone on the flanks of the upper Mavo Valley. Both areas showed airborne radiometric anomalies, but the results of soil sampling for bauxite were not encouraging, also the areas were too dissected to allow for the accumulation of a sufficient volume of residual mantle.

CRA Exploration Pty Ltd prospected the Nggae area of north-west Guadalcanal for bauxite (Rebek and Thomas, 1973). Six holes were augered on areas of gently inclined remnant volcanic apron. No sample attained commercial bauxite grade, and the areas of potential interest were too limited to justify further prospecting.

Iron

An American Army report of siliceous hematite and limonite in the area of Oa was followed up by Connolly (1960). The 'Cape Henslow Iron Formation', outcropping on the ridge that runs south-west from Marau Peak towards Oa village, was discovered to be essentially float and *in situ* basic volcanics, highly lateritised with much earthy red haematite and yellow limonite. It is likely to assay less than 30 per cent metallic iron, and is nowhere as thick as the 1000 feet (305 m) quoted in the original report.

Beach-sand prospects

Grover (1960) reported on investigations by J. W. Connolly on the beach sands of Guadalcanal, made in 1956–57 on behalf of East Coast Minerals Pty Ltd (Sydney) as part of a regional search for rutile prospects. There was no likelihood, however, of discovering payable deposits of rutile, zircon or monazite due to the absence of requisite source rocks on the island. Forty samples were collected: most concentrates proved to be primarily magnetite, with minor amounts of ilmenite and other iron ores. Traces of zircon were discovered in the sand of Wanderer Bay, and rutile was found in small amounts in Tiaro Bay and on the Tasimboko coast.

Hammond and Goss (1972) reconnoitred the beach sands of western Guadalcanal with a view to investigating the potential for concentrations of magnetite, ilmenite and chromite. Tiaro Bay and Wanderer Bay were visited and six samples were taken from the sand of Beaufort Bay: heavy minerals yield for the latter ranged from 11.1 to 82.3 per cent. No economic potential was apparent.

Pisarski (1968) collected beach sediments from the mouths of the principal Guadalcanal rivers as part of the 1965–68 UNDP Aerial Geophysical Surveys Project. He sampled beach sand on both sides of each river mouth just below the high water mark: heavy fractions of the samples were analysed for Cu, Zn, Ni, Co, Cr, Ag and Au. His report relates primarily to the use of beach sediments in geochemical prospecting rather than the assessment of heavy minerals.

Cementstone prospects

The calcareous sediments in the Honiara area have been sampled and analysed for consideration of their possible use in a cement industry. Eleven chemical analyses of calcareous arenites from the Honiara Beds and the adjacent Lungga Beds are presented in Hackman (1977). The CaO content ranges from 4 to 27 per cent; none of these samples could be considered suitable material for the manufacture of cement without admixture of calcium carbonate, which could be obtained in a fairly pure form either from the Mbonehe Limestone or from the Reef Limestones overlying the Honiara Beds. A complex of coralline limestone and calcareous arenites is reasonably accessible on the left side of the lower Lungga Valley 1 km south of Mbetikama Mission. The necessary materials are also available in the Mount Austen area.

Grover (1965) referred to a large deposit of unctuous ash found by Pudsey-Dawson and Thompson (1958) in the Tanaemba River in north-west Guadalcanal. During tests by the Mineral Resources Division a sample was fired at 950 °C, converting to a suitable brick material. A partial analysis gave 54.24 per cent SiO_2 and 5.73 per cent CaO.

Summary

Pre-Oligocene mineralisation in the Mbirao Volcanics, essentially a sporadic segregation and remobilisation of metallic sulphides in a mafic igneous or ophiolitic suite, was succeeded by a sequence of 'porphyry copper' and gold mineralisations from the Oligocene through to the Pleistocene.

Titley (1975) relates the episodic porphyry copper mineralisation in the Solomons to variations in convergence rates of the Australian and Pacific Plates. However, Taylor (1976) challenges this interpretation, as he considers that the regional tectonics indicate a predominantly tensional regime: in this reference Taylor is citing evidence from the Florida Islands, 30 km to the north of Guadalcanal. He emphasises that the massive sulphide deposits of marine mafic volcanic association, comparable with those of Cyprus, may be considered only as a 'minor localised facet of what is otherwise a monotonous pile of mafic volcanics'.

FIVE

Geophysical investigations

Gravity

History of research

Observations of perturbations of the orbits of artificial satellites suggested that the Solomons area was associated with a marked geoidal positive gravity anomaly; this aroused a great deal of interest amongst geophysicists, which has led since 1963 to the financing of a series of gravimetric expeditions to the Solomons (Grover, 1968a).

Present knowledge of the gravity pattern in the Solomon Islands derives from three operational phases:

1   The first gravimetric traverses on Guadalcanal were made in the north-central Plains area in 1960. Coleman and Day (1965) presented a map of the Bouguer gravity anomalies of the area between Honiara and Aola Bay, extending in part nearly 20 km inland as far as Gold Ridge.

2   The results of the 1963–64 land gravity survey of the Solomon Islands were presented by Laudon (1968); Grover (1968b) discussed some geological implications of the gravity data in the same volume.

3   In 1965 marine gravity surveys were undertaken in the Solomons region by the Hawaii Institute of Geophysics. The results have more bearing on the regional structure of the Solomons than on that of Guadalcanal in particular (Rose and others, 1968).

The regional picture

The Solomon Islands regional Bouguer anomaly map of Laudon (1968) shows that Guadalcanal coincides with a small regional high superimposed on the south-western flank of a broad low centred on and parallel to the axis of the double chain; a negative anomaly is associated with Indispensable Strait, and Guadalcanal lies on the upward gradient to the south.

The map of gravity trends (Figure 23) shows how the relatively negative zones coincide with the two axial basins, the 'Slot' and Indispensable Strait. From a preliminary analysis of sparker profiles, Rose and others (1968) estimate a thickness of more than 2 km of sediment in the 'Slot'. The steepest gravity gradients are found on the southern flanks of the Indispensable Strait Basin, leading up to the highest

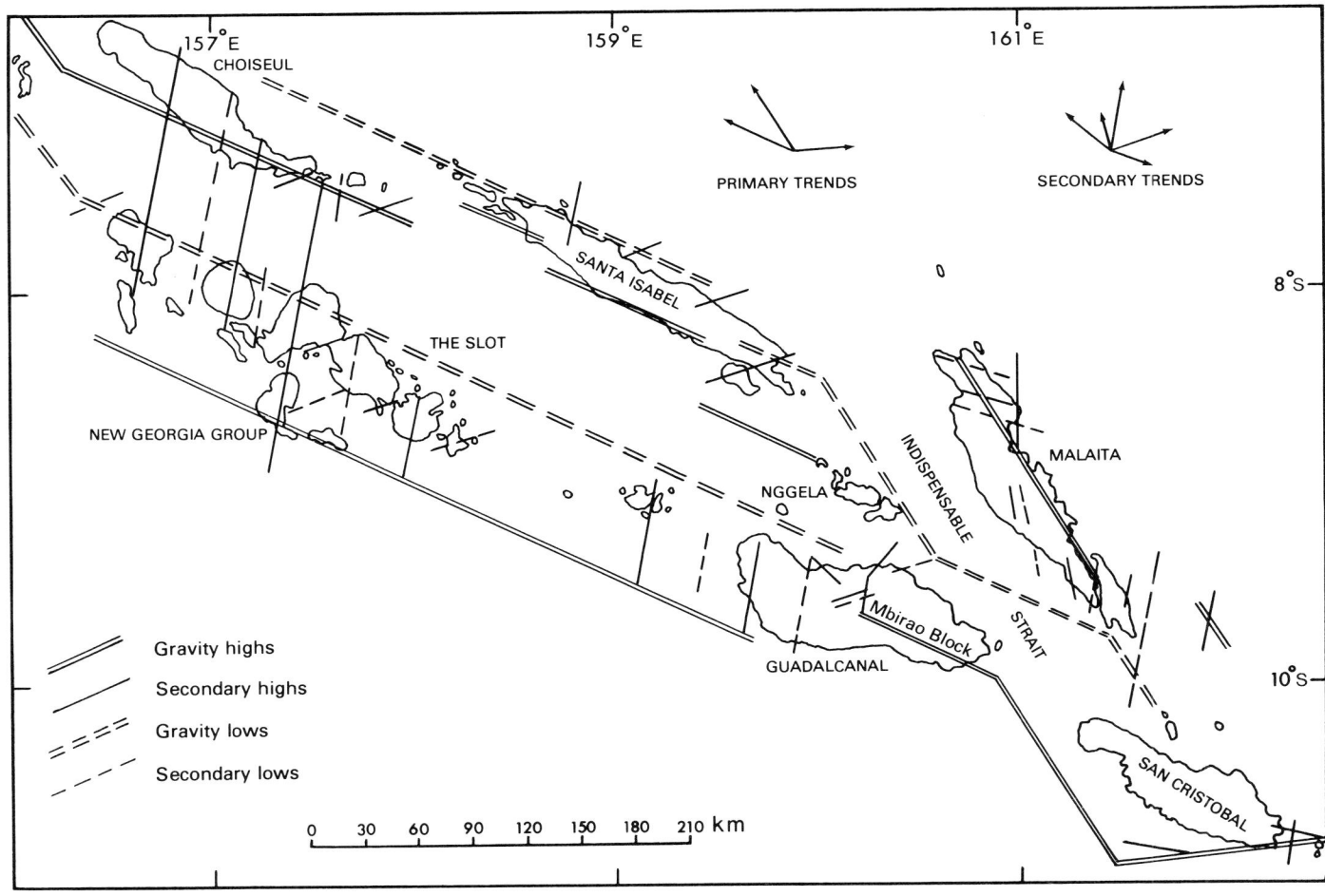

Figure 23   Map showing the trends of gravity anomalies in the Solomon Islands

positive readings (+ 250 milligals) in the south-east of San Cristobal.

Winkler (1968b) describes the regional gravimetric picture in the following terms: 'The Solomon festoon of islands occupies an arcuate low along the north-eastern flank of a tremendous Bouguer positive extending from the south-western Pacific over the Coral Sea to the edge of the Australian continental shelf.' However, from the configuration of the gravity contours the low might be regarded as roughly linear rather than arcuate in shape. In the central Solomons the calculated depth of the mantle/crust interface, according to Rose and others (1968), appears to vary from about 27 km under Indispensable Strait to 17 km south-west of Guadalcanal; from five other profiles across the entire Solomons–northern New Hebrides region, the estimated crustal thickness varies between 9 and 29 km. The deeper estimates for the mantle/crust interface are associated with the Slot and Indispensable Strait, with a bias towards the north-eastern side. Rose and others (1968) conclude that 'the Solomon Islands appear to have little or no "root" and have essentially the same depth to the mantle as the oceanic area to the north. Either the mode of origin of these islands is significantly different from that of most island-arc systems, or the tectonic activity is too recent to have permitted a "root" to develop'.

Although Laudon (1968) suggests that there has been regional isostatic compensation of the Solomons as a whole, there are significant departures from isostasy as far as individual islands are concerned, particularly in the New Georgia Group; according to Winkler (1968b) the analysis of regional data by the University of Hawaii showed that a 'large portion of the mass-inhomogeneity is of near-surface origin'.

The gravity pattern in Guadalcanal

The earlier work in north-central Guadalcanal revealed the following salient features (Coleman and Day, 1965):
1   Bouguer anomaly values increased inland from about — 10 milligals around Aola Bay to + 120 milligals at Gold Ridge.
2   The 'Tetere Anomaly' forms a distinct positive feature trending north-east from the Chovohio Valley towards Nughu Island (Figure 24).
3   Unusually steep gravity gradients occur on the flanks of the Tetere Anomaly: in particular that on the eastern side, in the area of the lower Mbokokimbo River, was the steepest positive gradient reported in geophysical literature at that time.

Figure 23 shows that the Indispensable Strait 'low' is separated from that of the Slot by a disturbance in the trend pattern associated with the Isabel–Nggela ridge. To the south, in Guadalcanal, this break in continuity is reinforced by a steep gravity gradient which corresponds broadly with the north-east-trending Mbalisuna–Malango fault zones. This line coincides with the eastern limit of Tertiary andesitic vulcanism in Guadalcanal (discounting the minor occurrence of Gallego-type lavas at Mbalo in the south-east).

Eight land gravimetric traverses were made across Guadalcanal (Laudon, 1968), giving a better control than for any other major island in the Solomons. The basic features of the gravity pattern (Figure 24) correlate remarkably well with what is known of the geology and structural domains:
1   The 'Mbirao Block' appears as a prominent high in the south and east, corresponding with anomaly values ranging from + 150 to + 210 milligals. It is bounded on the north and west flanks by steep gravity gradients. There is a clear correlation with the outcrop of the Mbirao Group and associated ultrabasics, but not with the Guadalcanal Gabbro to the west. Embayments in the isogal patterns suggest correlations with east-west lineaments (the Marau Ultrabasics) in the east, and with features trending north-east in central Guadalcanal (the Suta Ultrabasics).
2   Subsidiary highs in north and west Guadalcanal include:
a   The *Tetere Anomaly*. According to Coleman and Day (1965) the most probable explanation for this would be in terms of a buried ridge of basement rocks lying at relatively shallow depths under the sediments. Such a ridge should have acted as a kind of barrier between two basins of deposition, at least during the Miocene. This theory is supported by the following observations:
i   The Pliocene sediments in north-central Guadalcanal, which overlap onto the pre-Miocene in the Sutakama Valley are of paralic or shallow-water facies; these sediments overlie the Tetere Anomaly, and show rapid facies changes marking an area of transition between two basins of sedimentation.
ii   'Mbirao-type' metabasic rocks occur on the western flanks of the anomaly in the Tina and Mbetilonga headwaters; conceivably these rocks also form the floor to the Neogene sediments in the area of the Matepono headwaters.

Accordingly it is considered that the Tetere Anomaly is due to a step-faulted exclave of the Mbirao Block; however there may be other contributory factors at depth, for example, late Tertiary gabbroic intrusives may be present (they do in fact outcrop in the Lower Mbalisuna River); or Pliocene andesitic volcanics may be buried under the Toni conglomerates to the east of Gold Ridge.
b   The *Ghausava area*, a small positive anomaly with north-westerly trend, clearly associated with the Ghausava Ultrabasics and surrounding pre-Miocene basic volcanics.
c   The *Savulei area*, a less clearly defined 'high' in west Guadalcanal, corresponding with a belt of Miocene-Pliocene andesitic volcanics.
3   The *Lungga and Mbokokimbo 'shelves'*; these are relatively negative areas, corresponding broadly with Neogene sedimentary basins and separated by the Tetere Anomaly.

On the *Lungga 'Shelf'* the north–south regional gravity gradient is much gentler than to the east. The andesitic volcanics along the south coast and in the north-west of Guadalcanal do not show any pattern distinct from that of the sediments of the Lungga and Itina basins: it must be assumed that the denser materials characteristic of the Mbirao Group are either absent in this area or else occur at much greater depths than to the west.

The *Mbokokimbo 'Shelf'* corresponds broadly with the outcrop of the Mbokokimbo Formation, and is bounded on the south by the Mbirao Block, to the north-west by the Tetere Anomaly, and to the north-east by a very steep gravity gradient towards the Indispensable Strait Basin.

functioning at full efficiency on a regular basis. Where the trenches are clearly defined in the bathymetry the seismicity shows a distinct zonation; where the bathymetry is more complex, the seismic pattern is not so easily related to topographic features (Denham, 1969).

Three distinct seismic belts, corresponding to shallow, intermediate and deep hypocentres are broadly related to the Planet Trench where it is clearly defined on the south side of Bougainville; the pattern suggests that the seismic zone dips to the north or north-east in the manner of a semi-conical structure with its apex at depth north of Bougainville.

The pattern in the Santa Cruz group, which may be considered as an extension of the northern New Hebrides, shows

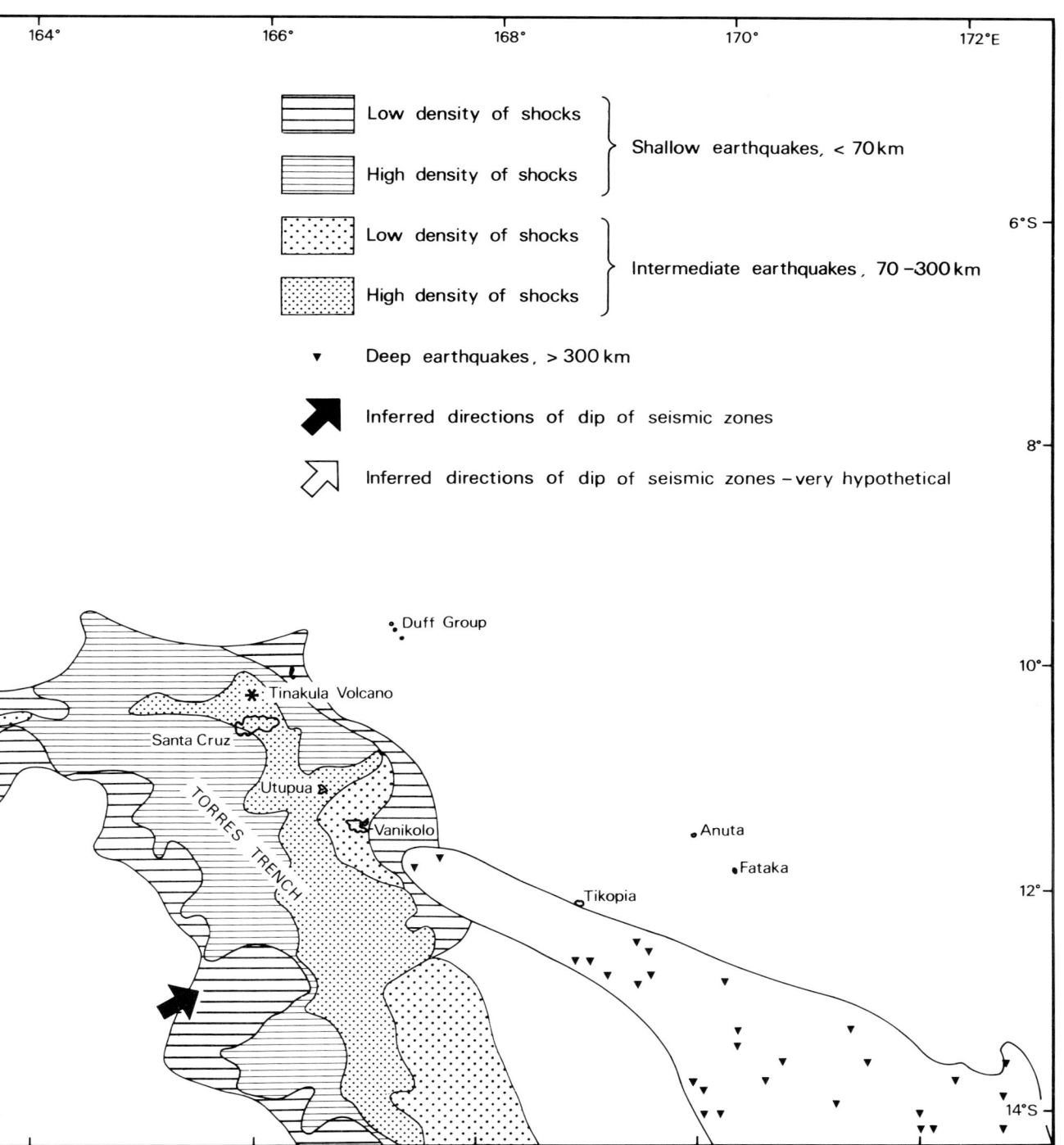

This 'shelf' might be compared to a step on the northern flanks of the Guadalcanal horst; its relationships with the structures to the north and south must have remained relatively stable throughout most of the Tertiary to allow for the deposition of a considerable thickness of fine-grained arenites.

Isostatic compensation

The gravity high of the Mbirao Block correlates very well with the outcrop of the Mbirao Group and with the main mountain range, although the highest mountains of Guadalcanal, including the Kavo Range and Haiacha group straddle the boundary gradient to the west. It follows that there is close correspondence between the gravity pattern and the 'basement' geology, but not necessarily with the topography. The younger volcanic piles of western Guadalcanal might be expected to show some lack of isostatic compensation comparable with that of the New Georgia Group, as interpreted by Laudon (1968).

Gravity trends

The gravity trends on Figure 23 were derived by Winkler (written communication, 1968) from the composite gravity map of the Solomons. Comparison with the trends of major structures on Guadalcanal reveals the following features:

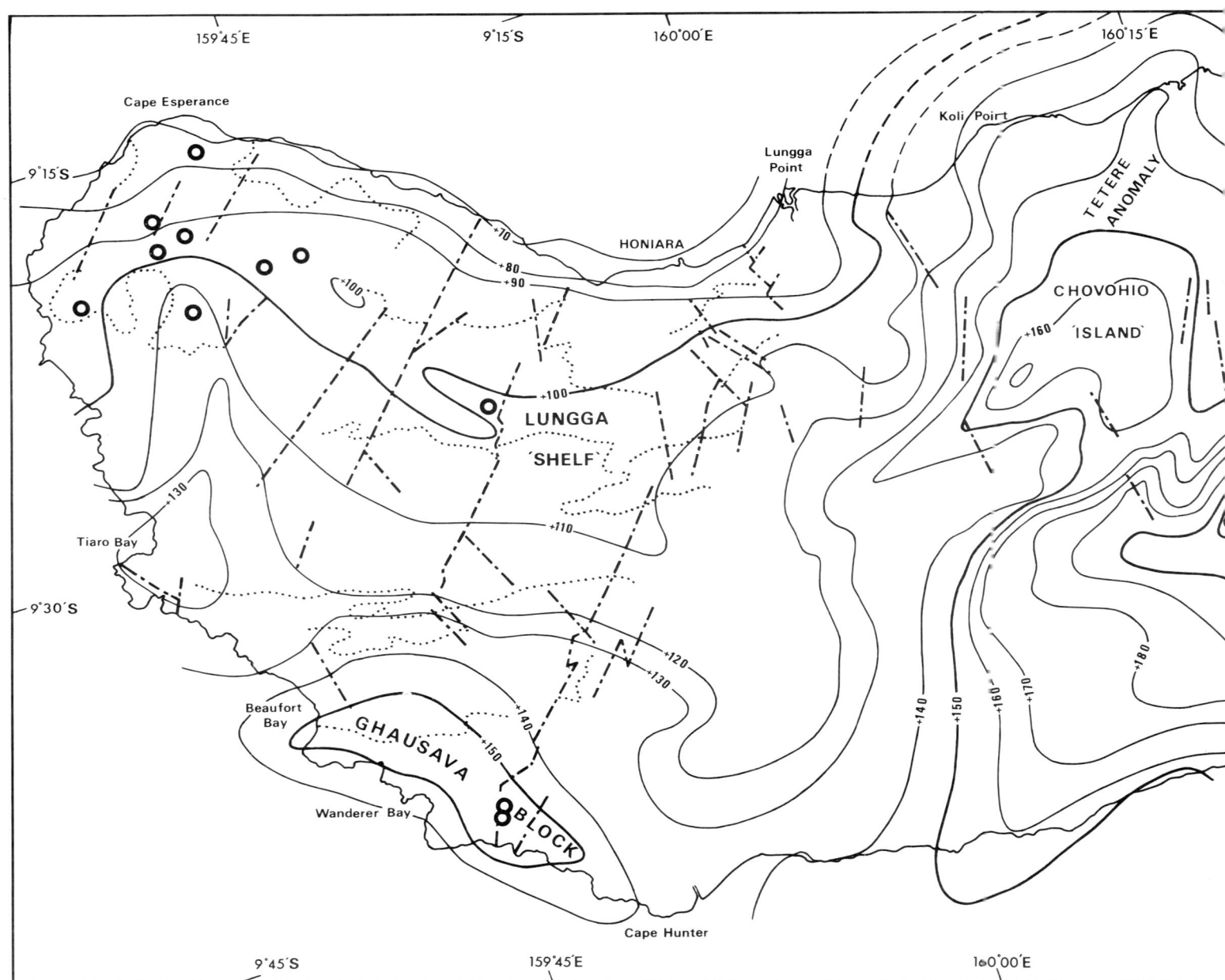

Figure 24 Map showing Bouguer anomalies, and geophysical units and discontinuities deduced from magnetic, electromagnetic and gamma-ray spectrometric data

1   Between Choiseul and eastern Guadalcanal north-westerly trends are dominant, being reflected particularly well in the linear outline of Santa Isabel; these trends parallel the direction of Recent sinistral transcurrent movement on Guadalcanal and are normal to the derived direction of maximum regional stress for the early Miocene.

2   In the Malaita–San Cristobal area, but also in Bougainville, the north-westerly 'grain' is broken by north-north-westerly trends parallel to the axis of Malaita Island. Comparable structures in Guadalcanal are the axes of the folds which affect Lower–Middle Miocene sediments.

3   North-north-easterly trends, although not persistent, transect the island blocks along the whole length of the

double chain. In Guadalcanal they manifest themselves as secondary horsts, predominantly Pliocene, whose axes are transverse to the dominant regional trends.

Conclusions

An excellent correlation can be made between the gravity pattern on Guadalcanal and the geological map. Positive anomalies correspond with uplifted areas of pre-Miocene basic–ultrabasic complexes, negative anomalies with Tertiary or Recent sedimentary basins.

The north–south regional gravity gradient is diversified by a series of steps, which indicate that large-scale block

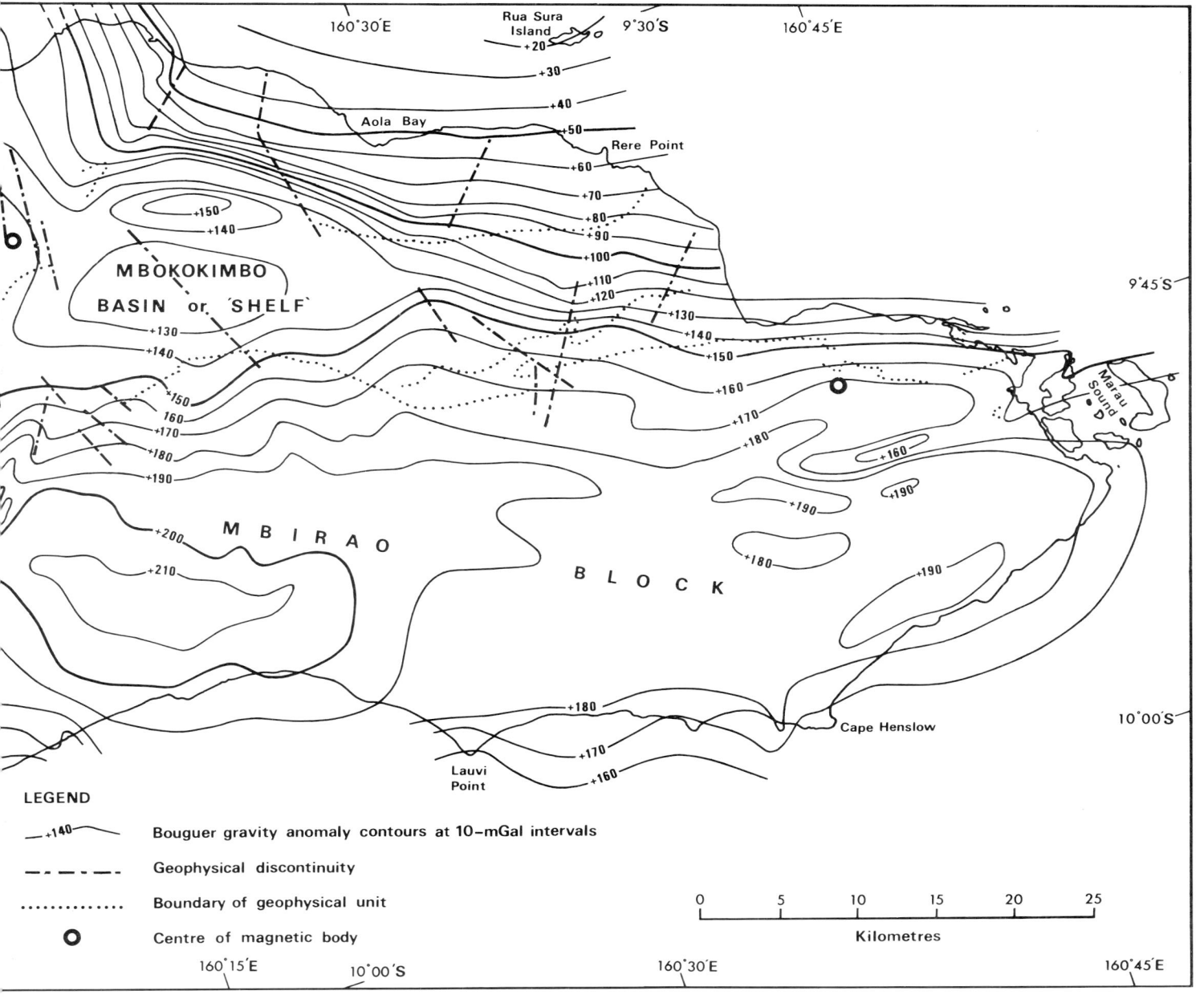

LEGEND

—+140— Bouguer gravity anomaly contours at 10–mGal intervals

— · — · — Geophysical discontinuity

· · · · · · · · · Boundary of geophysical unit

O Centre of magnetic body

faulting of the basement has played an important part in controlling the configuration of the sedimentary basins. Regional gravity trends tend to parallel axes of anticlinal horsting.

The relationships between topography and gravity anomalies suggest that the Mbirao Block, in particular, is not isostatically compensated.

Seismicity

Figure 25 shows a generalised seismic map of the Solomons–Santa Cruz Region; this is based on positions of hypocentres recorded by the United States Coastal and Geodetic Survey World Standard Network over the period January 1963–April 1969, that is, since the Honiara Observatory has been

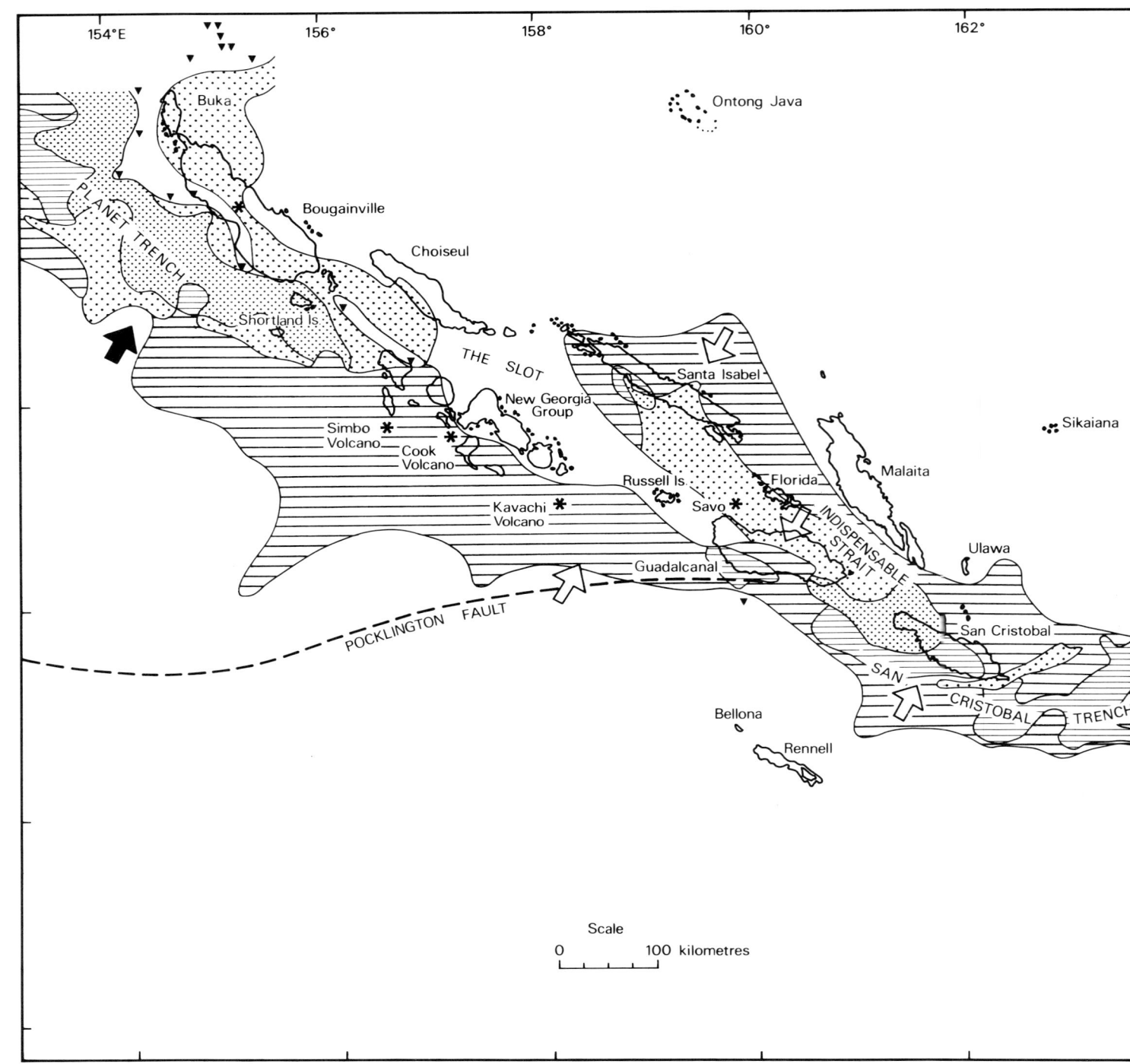

Figure 25 Map of seismic activity in the Solomons–Santa Cruz region

a comparable symmetry with deep focus shocks to the east and zones of intermediate and shallow shocks parallel to the Torres Trench.

The seismic pattern in between the Shortlands and San Cristobal is more difficult to interpret. With the exception of the western half of San Cristobal and the strait separating it from Guadalcanal, the density of shallow and intermediate hypocentres is much lower in the central Solomons than in the Santa Cruz and Bougainville areas. A zone of shallow shocks corresponds with the Recent volcanic belt on southern flank of the New Georgia Group and extends along the south coast of Guadalcanal, following the line of the San Cristobal Trench. Intermediate shocks cluster around the west end of San Cristobal, and are sparsely distributed in the Slot south of Santa Isabel. There is a poorly defined zone of shallow shocks to the north of the Santa Isabel–Florida ridge.

In the central Solomons only one deep-focus shock has been recorded in the trench south of Guadalcanal. This predominance of shallow shocks in the centre led Grover (1967) to interpret the Solomons in terms of an 'Arch of Seismicity'; thus a longitudinal profile including the Santa Cruz Group shows a progression from a zone of deep foci at the extremities to shallow shocks in the centre. This concept was linked with a postulated 'energy transmission phenomenon' in the upper mantle which purported to explain an ascending association in time between deep, intermediate and shallow earthquakes. It is not appropriate to discuss the validity of Grover's argument here, but the arch concept is of value in so far as it expresses the inferred variation in crustal structure along the line of the island arc, as well as across it.

Westwood (1970) prefers to interpret the seismic zone in the central Solomons as essentially vertical; he considers that, as a very hypothetical alternative, the seismic zone could be regarded as dipping at a low angle to the north under western San Cristobal, and in the case of Guadalcanal, steeply to the south (if the deep-focus quake in the San Cristobal Trench is taken into account); from the offsetting of seismic zones, such a configuration would require a dextral transform fault in between Guadalcanal and San Cristobal.

Clearly in the central Solomons the seismic evidence is still tenuous and too contradictory to enable a clear-cut solution to emerge. Concerning a possible contribution to the neotectonic picture the salient features, as far as Guadalcanal is concerned, are:
1 A zone of shallow-focus earthquakes along the south coast of Guadalcanal.
2 A cluster of high-density shallow to intermediate seismicity between Guadalcanal and San Cristobal.
3 A low-density zone of intermediate foci in the Slot–Indispensable Strait area to the north.

Discussing the tectonics of the Kinki area of south-west Japan, Hizita (1969) notes that the earthquake foci tend to 'concentrate along the marginal zones of rather rigid masses among the basement rock bodies and along the strike-slip faults'; comparing with the situation in the central Solomons the irregular seismic pattern cannot be explained in terms of a classic 'Benioff' Zone.

According to Hodgson (1958) fault-plane solutions for earthquakes indicated a predominance of strike-slip faulting in island arcs: he cited an example indicating sinistral shift along north-westerly lines in the Bougainville area, and dextral south-easterly components in the New Hebrides. Isacks and others (1968) consider that Hodgson's solutions were based on unreliable determinations of focal mechanisms and that in fact recent data indicate that dip-slip mechanisms are predominant in island arcs, due to the concentration of activity near the margins of mobile lithosphere plates, in zones of convergence: they do not exclude strike-slip mechanisms in zones of moderate and shallow activity. Until earthquake mechanism studies on the Solomons area in particular give specific information on slip vectors, correlations between neotectonics and seismicity remain very speculative.

More recent data on focal mechanisms are discussed in Ripper (1970) and Curtis (1973a, b).

Magnetic survey

An airborne magnetometer survey was an integral part of the UNDP Aerial Geophysical Surveys Projects; specific references to the results for Guadalcanal are given in the ABEM Company report (1967) and Winkler (1968b). A quarter of the island was excluded from the airborne survey, mostly the rugged pre-Miocene basement area.

According to Winkler (1968b) 'the normal visible evidence for subjective correlation between magnetics and geology is missing from the magnetic maps'. This impression could be in part due to the low apparent susceptibility contrast (Rose and others, 1968) between the anomalous intrusive source bodies and the 'country rock' in a complex which is essentially basaltic to andesitic in composition. The latter authors consider, on the basis of preliminary interpretations of magnetic profiles derived from traverses across the surrounding seas, that the majority of magnetic source bodies within the Solomons region are located in the upper crust and originate from intrusive or extrusive ultrabasic and basic rocks.

However, as far as Guadalcanal is concerned, the ABEM report suggests several correlations between geology and magnetics:

In general, the Mbirao Group correlates with a magnetic high, the sedimentary area to the north with a magnetic low, the boundary between the two areas corresponding roughly with the 41 500-gamma contour line.

The axes of the magnetic anomalies in the northern half of Guadalcanal show a distinct east–west orientation; however the magnetic relief of the area of Plio-Pleistocene Gallego Volcanics of north-west Guadalcanal is very high in contrast with the surrounding sedimentary areas; here the anomaly pattern does not conform to the east–west 'grain' typical of the rest of the island. The magnetic pattern of north-central Guadalcanal is complicated, reflecting the intense faulting which has been confirmed on the ground in the Mbetilonga Basin and the area around Gold Ridge. Intensive faulting is also indicated in the area between the Mbalisuna and Mberande rivers, extending to the northeast under the line of the Gharangi Knoll; here the indications are that various magnetic structures have been superimposed on an earlier pattern: ABEM (1967) suggests that the pattern reflects intrusion of basic material into the fault zone. The discovery of the Mbalisuna Gabbro in the lower part of the Mbalisuna River certainly confirms this interpretation.

To the east the Valasi Block corresponds very well with a magnetic high; to the north-east, in the area behind Rere Point, the magnetics reflect the northward extension of basement structures below the sedimentary pile; the indications are that the sediments decrease rapidly in thickness from Aola eastwards to Kaoka Bay.

In eastern Guadalcanal the magnetic axes trend west-north-west, parallel to the outcrop of the Marau Ultrabasics.

This trend is maintained, following the Valasi Fault Complex as far west as the Mongga River; here the magnetic axes are deflected to a predominantly north-east trend in north-central Guadalcanal, parallel to the Suta Ultrabasics; it may be inferred that the magnetic axes reflect the major trends in the basement, and that the swing in strike of the ultrabasics from Marau through to Suta is completed at depth under the sediments of the Mbokokimbo Basin.

Lines of geophysical discontinuity

The lines of geophysical discontinuity shown on Figure 24 have been interpreted as deep-seated major faults (ABEM, 1967); they demarcate zones of sharp transition between domains which are characterised by distinct geophysical patterns.

Discontinuities trending north-west occur under the sedimentary pile east of the Mbalisuna fault zone; these are parallel to the axes of folding in the Lower and Middle Miocene, and to a swarm of major faults which traverses the Mbirao Block. The Paripao Discontinuity, which underlies the Mbokokimbo Basin, could conceivably be regarded as a northwesterly extension of the Chimiu Fault Zone, possibly offset dextrally along a north-easterly fault during the second phase of tectonism. The same trend is prominent in the joint-frequency diagram for the area (Figure 19, area 10). In north-central Guadalcanal the discontinuities trend north-north-east, particularly near the line of the Mbalisuna Fault Zone; again it is interesting to note that they parallel the joint-frequency maximum for this area (Figure 19, area 9), and also parallel the axes of Pliocene anticlinal horsting which manifestly affected parts of central Guadalcanal.

The eastern end of the Marau Ultrabasics is outlined as a geophysical unit distinct from the surrounding basement rocks. However, west of a major discontinuity along the line of the lower Kolovaghamela River, the continuation of the northern boundary of the ultrabasics, as interpreted from geophysical data, is deflected 6 km to the north, still running parallel to the serpentinite body. It may be inferred that the serpentinite mass is disposed vertically in the crust east of the Kolovaghamela River, but to the west is inclined at a moderate angle to the north. The maxima for S_2, associated with the second phase of tectonism in both the ultrabasics and the surrounding metabasics, show a corresponding tendency: the northerly dip of the s-surfaces is steeper in the east than in the west.

This same geophysical discontinuity can be traced west-north-west along the line of the Valasi fault zone, to swing round under the Mbokokimbo Basin parallel to the local trend of the magnetic axes. Conceivably this could be regarded as linking up with the northern limit of the Mbirao Metabasics at depth, and coinciding with a zone of intermittent ultrabasic emplacement.

In western Guadalcanal the Vatunjae Discontinuity, trending north-north-east, is suggestive of a basement wrench fault, obscured by Plio-Pleistocene sediments; the disposition of the Ghausava and Itina ultrabasic bodies on either side of this line indicates offsetting due to left-lateral strike-slip.

SIX

Regional geology

Summary of geological history

In Guadalcanal the fundamental stratigraphic division is between the pre-Miocene Mbirao Group and the overlying Neogene–Quaternary complex of volcanic and sedimentary rocks. There are four significant contrasts:

1 The absence of terrigenous material in the Mbirao Group, and its abundance in the overlying sediments.
2 Petrochemically, the oceanic or alkali-basalt/tholeiitic character of the Mbirao Volcanics, contrasting with the tholeiitic and incipient calc-alkaline trends of the younger volcanics.
3 The rapid horizontal changes in lithofacies within the post-Oligocene sequence, and the apparent relative homogeneity of the Mbirao Group.
4 The widespread evidence for tectonism and metamorphism, with the development of penetrative planar fabrics within the Mbirao Group, in contrast to the relative simplicity of the post-Oligocene tectonics, with its emphasis on normal faulting and only minor gentle folding.

With these differences in mind, the geological history of Guadalcanal may be appropriately divided into three major phases: the oceanic phase, the tectonic phase and the volcanic or fractured island arc phase. Each of these phases is characterised by geological features which bear fruitful comparison with significant aspects of the three major 'Provinces' of Coleman (1965a) (Figure 3). These are: the *Pacific Province* (Malaita and the northern fringe of Santa Isabel), predominantly an oceanic phase; the *Central Province*, in which the intense fracturing, metamorphism and ultrabasic intrusives are also the essential features of the tectonic phase in Guadalcanal; and the *Volcanic Province*, exemplified by the New Georgia Group, in which a pattern of emergent volcanic islands with fringing and off-shore reefs provides an environmental framework for the accumulation of varied sediments of volcaniclastic and biogenic origin.

Adjacent phases may overlap in time; thus in Guadalcanal features of the tectonic phase continue into the younger volcanic phase, and in Malaita a modified tectonic phase is partly coeval with the oceanic phase. Essentially, however, the trilogy appears to hold good.

The oceanic phase

No absolute data are yet available on the age of the Mbirao Volcanics. Probably during the Cretaceous there was widespread submarine extrusion of basaltic lava, chemically of alkali olivine-basalt or 'oceanic tholeiite' type (Engel and Engel, 1964). The Mbirao Volcanics were extruded directly onto the ocean floor or, in part, under a thin cuirass of pelagic biomicrite, the Tetekanji Limestones. Conceivably the effusion of lava occurred at moderate depths, with correspondingly high hydrostatic pressures: widespread metasomatism ensued, with albitisation of plagioclase, and segregation of monomineralic bodies of calcite, epidote, quartz

and cryptocrystalline silica. Under increasing depth of burial, zeolitisation was effected at lower levels in the lava pile by the action of circulating hydrothermal solutions. Concordant bodies of cognate dolerite were emplaced, even at relatively high levels in the lava pile. The altered 'greenstone' aspect and the intercalated cherts are reminiscent of the spilitic association of orogenic regions; however the greywackes, which frequently accompany this association elsewhere, are absent from the Mbirao Group.

Thompson (1960) and Coleman (1965a), discussing the early history of the Central Province, favoured the concept of initial 'geanticlinal welts' in the oceanic floor, the crests of which yielded under tension to create fissure-type feeders through which basaltic magma was intermittently extruded. This raises the question of the source of the basaltic magma: the Guadalcanal Gabbro intrudes the lava pile in depth, but the one analysis available (Table 3) indicates that in comparison with the Mbirao Volcanics it is very rich in alumina and magnesia; other petrographic studies also indicate the predominance of a leucocratic type. Possibly both gabbro and basalts were derived from a tholeiitic magma, the gabbro representing a more feldspathic differentiate. Local intratelluric segregation, which on a small scale gave rise to the glomeroporphyritic dolerites, could have taken place on a larger scale within the Guadalcanal Gabbro, leading in the Upper Cretaceous to the emplacement of extensive bodies of leucogabbro at high levels, for example in the Valasi Block. Pyroxene was uralitised before the completion of cooling.

Assuming that gabbroic material formed the cores of the hypothetical geanticlinal welts, dykes in the already solidified Guadalcanal Gabbro could have acted as feeders during extrusion of the Mbirao Volcanics; there is however no clearly developed parallel swarm: in the Valasi Block the dykes trend north-east, but in central Guadalcanal they strike north-west. One of these directions could conceivably have paralleled the axis of the original geanticline, the other resulting from tension normal thereto.

Intrusive basalt in the Sigana Volcanics on Santa Isabel has been dated at 66 ± 3 Ma, that is Palaeocene, about 15 Ma older than the date of metamorphism of the Choiseul Schists (Snelling, written communication, 1969). A comparable date is tentatively suggested for the hypabyssal phase (Mbirao Dolerites) in the pre-Miocene basement of Guadalcanal.

The tectonic phase

Tectonism and great mobility characterised the Eocene–Oligocene; the core of the Mbirao Geanticline was metamorphosed, ultrabasics were emplaced into axial dislocations, and two major phases of tectonism each imprinted a set of metamorphic s-surfaces.

During the initial phase of uplift and metamorphism, the axis of maximum horizontal regional stress was orientated north-west to south-east in relation to present-day structures. During the course of evolution of the geanticlinal welt the uplift developed a marked asymmetry, with overthrusting at depth towards the south-east. The ultrabasic and associated metagabbroic bodies of Suta and Marau were emplaced preferentially into axial dislocations, their disposition conforming with that of the axial-plane schistosity (S_1). Possibly emplacement of the Ghausava–Itina Ultrabasics (Thompson, 1968) also began at this time.

Metamorphism was concentrated in the 'hinge' area of the short limb of the composite wedge structure, predominantly a greenschist facies (Turner 1968), although tectonic slivers of plagioclase-amphibolite associated with the ultrabasic emplacements suggest significant increases in metamorphic grade in depth.

North of the Marau–Suta Ultrabasics line metavolcanics of Mbirao type are preserved only in the Mbeambea Valley of the Tina river system and possibly in the Mbetilonga area; elsewhere Guadalcanal Gabbro appears to form the floor to the sedimentary succession throughout north-central and north-east Guadalcanal. Although the Mbirao Metabasics have in general been more thoroughly metamorphosed than the Guadalcanal Gabbro to the north and west, it may be inferred that, in so far as the gabbro represents a deeper structural level, the crest of the original anticlinal horst lay to the north of the ultrabasic belt.

If it be assumed that the base of the Mbirao Volcanics was, at least locally, devoid of great irregularities, a displacement of 2000 to 3000 m must have been effected along the line of the Suta–Marau Ultrabasics, the Guadalcanal Gabbro having been upthrust on the north side. Parallelism of the melanocratic schlieren in the Guadalcanal Gabbro with the Suta Ultrabasics and S_1 surfaces in the Mbirao Metabasics indicates that the gabbro was still mobile in the earliest phases of tectonism during which the ultrabasic bodies were introduced into the major Suta–Marau dislocation. The intricacy of the serpentinite–gabbro boundaries reinforces the impression of mobility at depth.

Possibly the main phase of metamorphism, associated with the development of S_1 surfaces, was broadly concomitant with that of the Choiseul Schists, that is Middle Eocene (Richards and others, 1966). Monomineralic/oligomineralic segregations within the Mbirao Volcanics were mobilised during metamorphism, giving rise to extensive quartz-epidote venation; syngenetic metallic sulphide, particularly pyrite, segregated preferentially on S_1 surfaces: local segregations of magnetite and associated sulphides were diverted into minor lodes.

With continued tectonism, the southern limb of the asymmetrical uplift may have been obliterated; certainly no evidence for an inverted limb has been detected within the Mbirao Metabasics, although this failure may be due simply to the difficulty of identifying primary 'way-up' structures within the metamorphosed volcanic pile.

The second phase of tectonism made a stronger impression on eastern than on central Guadalcanal; a primary regional stress from the north-north-east profoundly modified the configuration of the Marau Ultrabasics, whose sigmoid outline may be interpreted as a compromise between trends related to S_1 and S_2; during the latter phase some adjustment was achieved by dextral wrenching on north-westerly lines. Displacement tended to be directed preferentially along established lines of weakness, such as were provided by the more continuous of the Tetekanji Limestone occurrences.

With the close of the second phase of tectonism, possibly during the early Oligocene, the Mbirao Block began to emerge above sea-level, essentially an anticlinal horst of oceanic lavas or tholeiitic 'root' material (Dewey, 1969).

The later volcanic or fractured island-arc phase

Possibly in the late Oligocene, but certainly extending into the Lower Miocene, differential block-faulting conditioned the emergence of islands of pre-Miocene material. After emergence of the Mbirao Block, possibly also of the Ghausava area of western Guadalcanal, rapid erosion led to the denudation of the metabasic zone in eastern Guadalcanal and the gabbro and ultrabasics to the north and west. A large volume of basaltic lava and gabbro must have been removed at this stage, the derived sediments having been since eroded or buried under the Neogene succession.

A second phase of vulcanicity of 'tholeiitic' or basaltic andesite type commenced, probably during the Oligocene in north-west Guadalcanal; basaltic andesites were extruded and diorite was emplaced at depth, for example in the Poha River area: this phase continued into the Lower Miocene in southern and south-western Guadalcanal. The Marasa and Suta Volcanics were extruded in part into fringing or offshore reef conditions in the Lower Miocene: pillow structures indicate sub-aqueous extrusion, although part of the Suta Volcanics appears to have been erupted sub-aerially. Pyroclastic activity was more in evidence than before. Faults trending north-north-east were activated, particularly in central and western Guadalcanal. At that time, western Guadalcanal must have been a complex of small offshore volcanic islands, separated from each other and from the emergent Mbirao Block by a pattern of shallow channels and deeper basins.

In south-western Guadalcanal the Suta Volcanics provided a dominant source area for sedimentation of volcanic wackes, notably the Kavo Greywacke Beds and the greywackes of the lower part of the succession in the Itina Basin; the Itina–Lungga trough formed a natural 'sink' which was not completely filled until the end of the Pliocene.

In the early Lower Miocene (approximately upper 'e' in the Indonesian letter-scheme) the area must have stabilised sufficiently to allow fringing reef to accumulate in central Guadalcanal, represented by the Mbetilonga Limestone. The early part of this Mbetilonga phase was concomitant with the eruption of the Suta Volcanics, which accounts for the admixture of a large proportion of lithic volcanic material to the biogenic limestone.

By the late Lower Miocene (approximately f1–2) fringing reef conditions had extended to north-west Guadalcanal (the Mbonehe Limestone) and to the east as far as Mt Vatupochau (the Lake Lee Calcarenite). At that time, however, eastern Guadalcanal formed a high positive area, including the Valasi Block: the land area probably extended well to the south of the limits of the present-day Weather Coast Block. Presumably the greater mobility of the area west of the line of the Suta Ultrabasics contributed to the more rapid erosion of the central areas as compared with the east.

Apart from the Mbonehe Limestone, the late Lower Miocene limestones have a higher proportion of clastic calcarenites than biostromal limestones, with considerable terrigenous admixture: this indicates higher-energy sedimentation involving the partial destruction of older reefs.

The Tangareso Beds and the sediments of the Mbokokimbo Lithosome are products of similar sedimentary environments. The poorly-sorted fine-grained arenites may certainly be in part of tuffaceous origin, but in general they represent relatively quiet conditions with water either too dirty or too deep to allow the accumulation of biogenic limestones. Steady turbid flow provided the finer silty debris, with occasional higher-energy phenomena, possibly initiated by tsunamis (Coleman, 1968), depositing the coarse sandy and paraconglomeratic beds. Some sedimentary structures in the Tangareso Beds indicate derivation from the north-east; they may be regarded as having been deposited in a channel between the Mbirao 'Island' and an offshore island or peninsula. The gravity pattern strongly suggests the presence of such a 'buried island' under the middle basin of the Chovohio River, that is the 'Chovohio Island' of Figure 24.

The 'Tangareso Channel' was obliterated by infilling early in the Pliocene, but the thickness and extent of the Mbokokimbo Lithosome indicate that a broad gulf, open to the ocean, remained a stable feature of the palaeogeography of east-central Guadalcanal from the Upper Miocene through to the Pleistocene. The gravity pattern indicates that the Mbokokimbo Basin is founded on a shelf marginal to the Mbirao Block: this must have subsided steadily from mid-Miocene to late Pliocene, remaining marginal to areas of paralic alluviation and receiving a continuous supply of fine-grained arenite; accessible to the ocean, the gulf received a steady rain of planktonic foraminifera.

Renewal of the tectonic phase

In north-central Guadalcanal in the Middle Miocene the Charikangge Grit, deposited in part as turbidites (Coleman and McTavish, 1964), heralded a period of instability and intense alluviation centred on the area of Gold Ridge on the flanks of the 'Chovohio Island'.

This phase of instability may be correlated with renewed anticlinal horsting which particularly affected the island west of the line of the Suta Ultrabasics and resulted in the uplift of the Malango and Ghausava axes. According to Thompson (1960, 1968), the Ghausava–Itina ultrabasic bodies, emplaced in the Oligocene, were probably thrust higher into the crust at this time. The horsts were flanked by major fractures, in part reverse faults, notably the Suta, Mbeambea and Ghausava Faults of central and western Guadalcanal. The Mbirao Metabasics and Guadalcanal Gabbro were tectonised along the line of the Mbeambea Fault, where they were locally thrust to the north over the Mbetilonga Limestone.

East of the line of the Suta Ultrabasics the unconformity at the base of the Lake Lee Calcarenite was tectonised locally in relation to faults trending north-west.

The composite stratigraphic section (Figure 4) demonstrates major irregularities in the upper surface of the Mbirao Group in central Guadalcanal, notably the Mbahomea and Suta 'ridges'. The Mbetilonga Limestone fails to outcrop over the Mbahomea Ridge, while in the case of the Suta Ridge the conglomerates of the Pliocene Toni Formation rest directly on the Suta Ultrabasics, so that the

Miocene is absent from this area; only the Tangareso Beds may be represented in very condensed form.

Faulting along north-easterly lines in central Guadalcanal gave rise to a miniature 'horst and graben' topography which provided a series of basins of deposition in the Lower Miocene, such as the Tina and Tinahulu basins. The Mbokokimbo Basin or 'Shelf' remained as a gulf until the Pleistocene, but west of the Suta Ridge the pattern of basins was more irregular and short-lived.

In the early Pliocene the Miocene sediments responded to reactivation of basement faults by folding along north-westerly axes; gentle open *Bruchfalten* formed in the Mbokokimbo and Itina basins; and over the crests of the Suta and Mbahomea 'ridges' certain Miocene formations were either not deposited or were rapidly eroded during evolution of the *Bruchfalten*.

During this phase sedimentation was probably more or less continuous in the Mbokokimbo Basin: penecontemporaneous deformation, reminiscent of the 'syntaphral tectonics' of Carey (1963) is a feature of the middle part of the Mbokokimbo Lithosome and some of the conglomerates in the lower part of the Mberande Tongue of the Toni Lithosome.

In the Mbirao Metabasics earlier schistosities were folded at this time, and S_3 and S_4 surfaces were imprinted on the basement rocks, being particularly well developed in south-central Guadalcanal and the extreme east.

The shaping of Guadalcanal

The effect of early Pliocene uplift and vulcanicity in western Guadalcanal was to unite the Oligocene–Miocene volcanics and sediments with the Mbirao Block to form one coherent land mass, leaving the Lungga Basin as an almost land-locked focus of sedimentation comparable with the present-day 'Blanche Channel' in the New Georgia Group.

In the early Pliocene a phase of predominantly submarine vulcanicity was centred on the flanks of the 'Chovohio Island': a large volume of brecciated basaltic andesite was extruded. Petrographically the lavas resembled the Suta Volcanics, although hypersthene-andesite was extruded in a later (probably Upper Pliocene) phase. Gold–quartz–pyrite mineralisation, associated with the later stages of vulcanicity, also affected the Sutakiki area to the south.

The Toni Beds represent a period of intense alluviation associated directly with the vulcanicity and tectonic instability of central Guadalcanal. The majority of phenoclasts in the conglomerates are clearly derived directly from the Gold Ridge Volcanics; basaltic flows were first brecciated, the large clasts rounded, and the material partly remobilised as submarine mud flows which deposited an extensive blanket of volcanically derived conglomeratic material on the sea floor around 'Chovohio Island'. The scope of the 'Toni Lithosome' embraces the paralic and shallow-marine environments of Tercier (1940); interpretation of the environment is blurred by the degree of tectonism and instability (Krumbein and Sloss, 1963, p. 428).

During the Pliocene the diverse pattern of emergent volcanic islands and intermittent tectonism and uplift led to rapid horizontal variations in lithofacies. In western Guadalcanal a steady supply of lithic and volcanic wackes continue to accumulate in the Itina–Lungga Basin.

In north-eastern Guadalcanal, however, quieter conditions prevailed at this time; fringing reef (the Valasi Limestone) was accumulating in the area of the Valasi Block, to form an easterly extension, but a much younger analogue, of the calcarenite–biostromal limestone facies of the Lake Lee Calcarenite.

Later in the Pliocene the energy level of sedimentation decreased generally throughout north-central and eastern Guadalcanal; the Mbetivatu Sandstone, and the Mbalanga Shale and Kolohaisava Mudstone 'Tongues' of the Mbokokimbo Lithosome are largely arenites of shallow-marine environment; sedimentary structures indicative of littoral or epineritic conditions are particularly evident within the Mbalanga Shale 'Tongue'; ecological evidence also indicates estuarine conditions in the Upper Toni Beds.

Renewed uplift in the Upper Pliocene is indicated by the increased energy of sedimentation, conglomerates again predominating in the upper part of the Toni Formation, the Paripao Tongue and the Vatumbulu Beds. Biogenic limestones accumulated locally on stable shelves sheltered from rapid alluviation. The Vatumbulu Beds were clearly derived directly from erosion of the Mbirao Block to the south: to the west, however, the Upper Pliocene conglomerates and arenites received an increasing supply of material derived from andesitic lavas of Gallego type.

In western Guadalcanal calc-alkaline magma was certainly available at depth early in the Pliocene as indicated by the age-determinations for the Koloula Diorite; hornblende-andesites were extruded over a large area of north-western Guadalcanal, and appeared even at the south-eastern extremity of the island: they provided a source of detritus for the clastic sediments of the Upper Pliocene and Pleistocene. The Mbalisuna Gabbro may also have been emplaced during this phase, indicating, somewhat anomalously, the availability of a more basic alkaline magma at a high level in the crust at about the same time that intermediate–acid lavas were being extruded in a closely neighbouring area.

Vulcanicity was linked with reactivation of faults, particularly north-easterly faults in the Savulei, Malango and Suta areas. It has been suggested that the Koloula Diorite was emplaced into a north-easterly zone of tension; the axes of minor late-Pliocene horst and graben structures indicate the same trend, for example the Tausoro Axis in north-central Guadalcanal.

By the Pleistocene Guadalcanal had almost assumed its modern shape. The Honiara Beds were forming as a northern fringe complex of reef and shallow marine or estuarine sediments. Early in the Pleistocene the sea extended up to 25 km inland, as far as the Haviha Ridge; marine erosion was short-lived, however, and the interaction of intermittent uplift with eustatic changes resulted in the magnificent series of terraces behind Honiara. The Suta Volcanics and derived Kavo Greywacke Beds, which had accumulated, in part, in a fairly deep Miocene trough, were now elevated to heights of over 2000 m above sea-level in the Kavo–Haiacha Ranges.

In the Holocene, the pronounced north–south geomorphologic asymmetry of Guadalcanal may be linked with the reinforcement of a geologic pattern: continued shallow-marine sedimentation and deltaic spreading on the northern coast, and rapid erosion and active tectonism on the south side. The profiles (Figure 6) strongly suggest that a large volume of the basic volcanic pile has been faulted down in step-like fashion towards the western end of the San Cristobal Trench. The recent offsetting of geomorphologic features along the Veuru Moli Fault is in line with this picture of continued tectonism.

Comparative regional geology

The Solomons in general and Guadalcanal in particular share many geological features in common with other areas of the Melanesian Complex or 'Outer Melanesian Arc' of Glaessner (1950). The Tertiary of Fiji, the New Hebrides and northern New Guinea is characterised, as in the Solomons, by:

1 A long record of vulcanicity, the extrusive products ranging from oceanic-type alkali-basalts to calc-alkaline andesites, and in New Guinea and Fiji, to a shoshonitic association (Jakeš and White, 1969; Joplin, 1968, p. 35). The development of the orogenic calc-alkaline trend appears to be a Neogene–Recent feature;

2 A great variety of sedimentary environments, characterised by rapid horizontal facies changes: the sediments are part of autochthonous stratigraphical systems (Coleman, 1966), being either derived from local igneous detritus (largely volcanigenic) or of biogenic origin. However the regional stratigraphic columns show a very real similarity throughout the area;

3 Evidence for more or less continuous tectonic instability, which has given rise to complex patterns of faulted anticlinal horsts or composite wedges. Changes in direction of regional trends as evidenced by 'sigmoid' effects suggest complex interaction of different stress regimes within sequences of polyphase deformation;

4 A complex pattern of Recent seismicity, with a zonation in relation to the Pacific Ocean which is the reverse of that shown by the other island-arc systems bordering the 'Andesite Line'.

In this section an effort is made to highlight some of the more obvious comparisons between the regional geology of Guadalcanal and that of other parts of the Solomons and Melanesia in general.

The pre-Miocene basement

A complex of essentially tholeiitic lavas, with subsidiary 'oceanic' limestones, intruded by gabbros and ultrabasics, and locally subjected to a low grade of regional metamorphism, is characteristic of Coleman's Central Province (Coleman, 1965a). In most areas, however, unmetamorphosed lavas have been regarded as younger than metamorphosed basement, whereas the writer is in favour of regarding the latter as a derivative of the former.

Pending more detailed investigations, the Mbirao Group may be compared, on the grounds of petrologic association and stratigraphic position, with the following named divisions in other islands:

Choiseul: Voza (Vosa) Lavas and Choiseul Schists (Coleman, 1960a).

Santa Isabel: Sigana Volcanics and partly metamorphosed undifferentiated basement (Stanton, 1961).

Florida: Basement of ultrabasics, gabbros and altered pillow lavas—Nggela Group (Thompson, 1958).

San Cristobal: Warahito Lavas and Ravo Limestone (Thompson and Pudsey-Dawson, 1958).

Comparison with the Alite Volcanics which form the basement of Malaita Island (Rickwood, 1957) falls short on some points; although the Fiu Lavas are petrographically similar to the Mbirao Volcanics, no ultrabasic intrusives have yet been found on Malaita, and the lavas have been folded with the overlying sediments, but not regionally metamorphosed (Maranzana, 1968). Detailed inter-insular correlation must await the results of further radiometric age determinations and the discovery of identifiable planktonic foraminifera in the intercalated 'pelagic' limestones. Shallow-water terrigenous sediment is conspicuously absent from the pre-Miocene 'basement' rocks of the Solomons.

In the Melanesian area east of the Solomons, the Wainimala Group of Viti Levu, Fiji (Rodda and Band, 1967), described as a complex of spilitic basalts, greenschists and sediments, appears worthy of further comparative investigation with the Mbirao Group; in the New Hebrides an areally limited occurrence of ultrabasics, associated amphibolites and altered basaltic pillow lavas at the southern end of Pentecost Island (Figure 2) has been described by Mallick (1969).

A very interesting comparison may be drawn between the pre-Miocene of Guadalcanal and the Ultramafic Belt of eastern Papua, described by Davies (1968). At much longer range, the association is similar to that of the Troodos Complex of Cyprus (Gass, 1968). In each case large volumes of magnesian gabbro associated with peridotites appear to have evolved under oceanic conditions, yet resemble Alpine 'ophiolitic complexes'. Thompson (1967) describes the Upper Cretaceous submarine volcanics of east Papua, associated with foliated red limestones and Eocene cherts; this 'oceanic crust' was arched in the Eocene and eventually thrust south-westwards over Mesozoic sediments as a preliminary to orogenic emergence in the Lower Miocene.

The origin of these 'oceanic' basement rocks thus bears on the problem of the composition of the upper mantle itself; gabbro must be an essential constituent of the upper mantle, if one considers the evidence from such ultramafic complexes as those of Cyprus and eastern Papua in the light of Dietz's (1963) interpretation of peridotite and related eugeosynclinal rocks as fragments of 'ocean rind' formed along mid-oceanic ridges.

The Suta Volcanics and Poha Diorite

It is difficult to draw direct comparisons between the Suta Volcanics and similar Oligocene–Miocene extrusives elsewhere in the Solomons because of the dearth of detailed petrographic and petrochemical data. This is unfortunate in view of their importance in Guadalcanal as an intermediate phase between the pre-Miocene basic volcanics and the Plio-Pleistocene andesites.

Petrographically and petrochemically, the Suta Volcanics are similar to the 'big feldspar' basaltic–andesitic suite of New Georgia (Stanton and Bell, 1969), which is considered to be of Pliocene age. In the Florida group the altered basement volcanics are overlain by a sequence of porphyritic basalts, succeeded by non-porphyritic pillow lavas of probable Upper Oligocene to early Lower Miocene age— the Soghonara Lavas (Coleman, 1965a); however, detailed petrographic data are not available.

A basaltic–andesitic magmatic episode is found in the Oligocene–Lower Miocene in many Melanesian areas, associated with dioritic stocks attended by sulphide mineralisation. For example, in Bougainville Blake (1968) describes orogenic calc-alkaline andesites in the Crown Prince Range area. In the New Hebrides, diorite in south-western Espiritu Santo has been dated as late Oligocene–early Miocene: in the same island the Lower–Middle Miocene Mamasa and Pelowou Volcanics, predominantly pyroxene-phyric and feldsparphyric basalts with autobrecciated flows and basaltic peperites, are petrographically similar to the Suta Volcanics (Robinson, 1969); in Fiji the Tholo Volcanics, which include plutonics ranging from olivine-gabbro to diorite with some associated sulphide and gold mineralisation, are of comparable age.

The Kavo Greywacke Beds

Greywacke-type associations resembling the sediments of the Itina Basin and Kavo Ranges of south-west Guadalcanal have been described from north-west Choiseul—the Middle and Upper Miocene Mole Formation (Coleman, 1965a); similar beds have also been observed in north-western Santa Isabel. The association indicates rapid accumulation in a marine trough, sediment being derived from an adjacent high volcanic terrain under conditions of tectonic instability; it may be compared with the now-classic volcanic wackes of the Aure Trough in Papua (Edwards, 1950).

Miocene calcareous sediments

The biogenic limestones and related calcarenites of the Lower Miocene provide the best-defined marker horizons in the Melanesian stratigraphic column. Comparable lithologies and faunas indicate that similar environmental conditions obtained not only within the Solomons, but also in many parts of the south-western Pacific from Indonesia through to Fiji. The Mbetilonga Group, with its contaminated biostromal limestones and interreefal calcarenites has its counterparts on Choiseul (the Kamangga Grit and Mt Vasu Limestone), the Florida Group (the late Lower Miocene Anuha Calcarenite) and San Cristobal (Harigha Conglomerates).

In the New Hebrides, Robinson (1969) describes algal foraminiferal biomicrite in the Lower Miocene of Espiritu Santo (the Peteao Limestone), and in Bougainville Tertiary 'e' beds include the massive organic Keriaka Limestone (Blake and Miezitis, 1967); further west, throughout New Guinea, calcarenites and biohermal limestones are persistent features of the Oligocene–Miocene. Thompson (1967) emphasises the importance of the Lower Miocene as 'a distinct turning point in the history of the Papuan Basin': the indications are that it was a period of orogenic emergence of an oceanic area along much of the Outer Melanesian Arc and that emergence was followed by a period of stability sufficient to allow for the accumulation of biohermal limestone.

Plio-Pleistocene sedimentary sequences

The exceptionally rapid horizontal facies variation in the Plio-Pleistocene of Guadalcanal precludes detailed correlation with other major islands; also the foraminiferal assemblages have proved less susceptible to subdivision than those of the Miocene sediments (Coleman, 1965b). The complex of emergent volcanic islands, reefs and inter-insular channels gives a picture of a rapidly changing environment, for which the New Georgia Group today provides a suitable model.

Volcanic rudites similar to those of the Toni Formation are found in other major islands, but are nowhere as voluminous as in north-central Guadalcanal.

Poorly sorted arenites rich in planktonic foraminifera, typical of the Mbokokimbo Formation, have parallels in the Pliocene Pemba Siltstones of western Choiseul (Coleman, 1965a); the calcilutites are comparable with some of the muddier parts of the Alite Limestones and Tomba Silts of the Malaita Group. Also a direct comparison could be drawn with the Wounpouko and Tawoli Calcarenites of north-west Espiritu Santo, interpreted by Robinson (1968) as open inter-reef deposits.

The Honiara Beds have their counterparts in those areas where stable Pleistocene shelves allowed the accumulation of reefs, for example the Nukiki Limestone of Choiseul (Coleman, 1965a) and the Sohano Limestone, an elevated reef complex in Buka and northern Bougainville (Blake and Miezitis, 1967). The Honiara terraces may be compared with the Plio-Pleistocene terrace sequences of Malekula, New Hebrides (Mitchell, 1969).

Plio-Pleistocene igneous activity

The Gallego Lavas may be compared with the basaltic andesites of Vella Lavella (Stanton and Bell, 1969) in the New Georgia Group, and the Maetambe and Kumboro Volcanics of Choiseul (Coleman, 1965a). Post-Miocene volcanics cover a comparatively larger area of Bougainville (Blake, 1968), and are considered to belong to the orogenic calc-alkaline suite, directly comparable with the Japanese 'hypersthenic series' of Kuno (1959); the evidence from central Guadalcanal indicates that a thick sequence of extrusives in the Gold Ridge area, comparable with many of the types described from Bougainville, has been eroded and/or buried by late-Pliocene paralic sedimentation.

Calc-alkaline vulcanicity with emplacement of high-level dioritic intrusives like those of Koloula, south Guadalcanal and Panguna, Bougainville, can be found in the New Britain Arc and on the northern side of east Papua (for example the Mt Lamington Lavas). It also dominates the Pliocene–Holocene volcanic province of the New Hebrides, which Warden (1967) describes as chemically calc-alkaline but mineralogically tholeiitic, and parts of Fiji—the Men-

drausutha Andesitic Group of Rodda and Band (1967). Jakes and White (1969) admit that magma-type zonation is not distinct within the Solomons Arc: any linear zonation comparable with Kuno's (1966) magma-type/seismic zone correlation for Japan is clearly much obscured by fracturing and profound changes in trend of the arc components.

Structural geology

To date detailed structural data comparable with what has been presented here for Guadalcanal have not been collated for any part of the Melanesian area. Thompson (1960) interpreted the cleavage planes of the schists in eastern Choiseul as gently folded about axes plunging north-west; this is an excellent parallel with the situation in the Mbirao Metabasics, but a more detailed survey is required for the entire area of the Choiseul Schists and Voza Lavas before a fair comparison can be drawn. In this respect it is worthy of note that the trend of the ultrabasics in eastern Choiseul is comparable with that of the Suta Ultrabasics of central Guadalcanal, in that it departs radically from the regional north-westerly alignment.

Folding of the Miocene sediments in Guadalcanal has its counterpart in the Vura Syncline of the Florida Group (Thompson, 1958) and the continuous 'cascade' style of folding in Malaita (Rickwood, 1957; Hackman, 1968b); throughout these three areas the fold axes trend generally north-west.

The structural history of New Guinea as outlined by Thompson (1967) evinces some interesting parallels with that of Guadalcanal:
1 Arching of the East Papuan Ultramafic Belt in the Eocene (compare Phase F_1 in the Mbirao Group);
2 Overthrusting towards the south-west in the Oligocene (compare Phase F_2);
3 'Orogenic emergence' in the Miocene;
4 Pliocene folding along north-westerly axes in northern New Guinea (compare Phases F_3-F_4).

The New Hebrides, as a younger analogue of the Solomons (Coleman, 1965b), appear to be structurally less complex: the 'sigmoid' effects, which are clearly indicated by the *en échelon* pattern of islands and the changes in major trends of ultrabasic bodies in the Solomons, are not in evidence within the New Hebrides–Santa Cruz arc, although there is an indication of a swing in trend of major structures in southern Espiritu Santo, possibly linked with a sinistral offset between that island and Malekula.

Mitchell and Warden (1971) emphasise the importance of differential movement of fault blocks and the absence of folding, low-angle thrusting and wrench faulting.

The following major trends in the structural history of Espiritu Santo (Robinson, 1968) compare broadly with events in Guadalcanal:
1 Taphrogenesis, faulting and uplift at the end of the Lower Miocene (compare F_3);
2 A mild orogeny in the early Pliocene (compare F_4);
3 The 'shaping of the present-day islands' in the early Pleistocene.

The fault patterns in Espiritu Santo Robinson, 1968) and Santa Cruz (Craig, *in press*) show maxima trending approximately north-north-west, west-north-west and, locally, north-north-east; this compares well with that of the post-Oligocene sediments in Guadalcanal (Figure 18, areas 9, 10). Possibly a similar regional stress system was operative in these areas.

The Solomons fractured arc

Theories on Melanesian geology in relation to global tectonics

Since 1960 much has been written on the geology and especially the geotectonics of the Pacific Ocean and its marginal island and mountain arcs. When attention is focused on the Melanesian region, a quotation from Menard (1964) is still apposite: 'Melanesia is structurally complex and little-known and not much can be said about it that is not conjectural'.

Several themes recur in the literature which illustrate the major trends in thinking about the Melanesian area:

1 The concept of a Melanesian Rise, trending south-eastwards from New Guinea to Fiji (Menard, 1964). Some authors have thought that the Melanesian area was originally 'continental', its nature having been presumably obscured by prolonged involvement in the highly mobile Pacific margin. However, Officer (1955) considered the crustal structure indicated that a succession of orogenic belts had been built out over an oceanic crust.

2 The occurrence of the main trench system on the Australian rather than the Pacific side of the chains, and the intriguing geophysical aberrations. Deflections of orbital satellites over the Melanesian area stimulated geophysicists to initiate land and sea gravity surveys in the Solomons area during the sixties; it has been shown that the Melanesian area corresponds with a major positive bulge of the geoid.

3 Tensional stress and 'taphrogenic' features; Thompson (1960) described the early development of the island arc in terms of a geanticline with longitudinal and transverse tensional faulting leading to vulcanicity and the formation of individual island blocks. Coleman (1965a) also considered that the larger islands began their Tertiary history as geanticlinal welts with associated tension features.

4 Vertical displacement; the evidence for differential vertical movements between sub-crustal blocks, particularly uplift in the axial zones of the major islands. In the case of Guadalcanal, Coleman (1965a) cited a variation in elevation of about 8000 m in the level of the upper boundary of the pre-Miocene basement, over a horizontal distance of only 30 km.

5 Horizontal displacement; Carey (1958) included the Melanesian area within his Tethyan Torsion Zone, inferring that the regional structure between Indonesia and Fiji was dominated by sinistral transcurrent movement, common to a global shear belt which could be traced westwards along the Alpine belt all the way to Gibraltar. Wilson (1959) cited the Solomons as an example of an *active primary 'fractured' arc*, of 'straight shape and irregular features', which, by definition, has evidence for a great transcurrent fault along its length. The 'Melanesian Border Plateau' (Figure 2), to the east of Santa Cruz, was described as a zone of sinistral crustal shearing by Fairbridge (1961). Amongst others, Krause (1962; 1965) has emphasised the

importance of sinistral wrench faulting in the Coral Sea between the Solomons and New Guinea, particularly along the Pocklington Fault (Figure 26): Moody (1966) interpreted the Melanesian area as a segment of sinistral shear within a 'Primeval Equatorial Shear System'.

Glaessner (1950) drew attention to sigmoid trends in the structure of New Guinea, features comparable, on a smaller scale, to the 'oroclines' which Carey (1958) integrated into his global picture of transcurrent movement.

6 Continental Drift, in its various elaborated forms; for example:

a Coleman's (1967) hypothesis of the 'Melanesian Re-entrant' as a complex of island arcs which broke away in tension from the eastern side of the Australian continent.

b Variations of drift theory invoking different sub-crustal convection patterns to explain the pattern of ocean rises and peripheral trenches; Wright (1966) and Cullen (1967) both attempted to analyse the south-west Pacific on these lines.

c Plate Tectonics, discussed below.

The evolution of geanticlinal welts

The structural analysis of Guadalcanal has revealed the contribution of both vertical and horizontal movements; the former are most evident to the field geologist, the existence of the latter has been deduced from broad regional considerations by the 'geodynamicists', but is also indicated from localised evidence for strike-slip dislocation.

Figure 27 presents a generalised picture of the evolution of the geanticlinal welts as seen in section across the Solomons Arc:

Stage 1 shows the extrusion of submarine basalts which form a thin crust over gabbroic 'upper mantle' material.

In Stage 2 the oceanic crust is warped into a series of geanticlinal welts or 'whaleback swells'. Longitudinal tension fractures develop parallel to the axes of the swells and persistent vertical flow conditions the uplift of a horst-like structure. Concordant cognate intrusives are emplaced high in the basic lava pile; possibly ultrabasic material is available in the core of the geanticline at this stage, sustaining increase in volume due to serpentinisation by juvenile waters.

In Stage 3 a tendency to lateral sliding in the upper parts of the uplifted block gives rise to local overthrusting, as has been found in Guadalcanal, for example, along the lines of the Mbeambea and Mborambora Thrusts. The horsts become fan-shaped (compare Beloussov, 1962, p. 522).

Stage 4 shows how overthrusting may develop preferentially on one flank of the horst giving rise to a form resembling an asymmetrical 'composite wedge' (De Sitter, 1956, p. 249). At the same time low-grade dynamothermal metamorphism of the base of the lava pile and the underlying gabbro is concentrated in the hinge area, imposing a penetrant schistosity in the axial zone flanking the ultrabasic body. After serpentinisation the ultrabasics are intermittently remobilised. At a later stage the crest of the horst may collapse to produce a type of 'keystone graben' between a conjugate pair of asymmetrical wedges—compare the Owl Creek Mountains of Wyoming (Wise, 1963).

In eastern and central Guadalcanal the Marau and Suta Ultrabasics dip towards the north, whereas in western

Guadalcanal (Ghausava area) and in Santa Isabel the ultrabasics are inclined towards the south (Thompson, 1960) so that the asymmetry of the geanticlinal zone is reversed. Significantly the change in disposition between the Suta and Ghausava bodies is marked by a major line of discontinuity (perhaps a fossil transform fault—see Wilson, 1965) which trends north-east, following the swing in regional trends and coinciding with the Suta–Koloula–Gold Ridge zone of post-Oligocene vulcanicity.

In Stage 4 the submarine ridge has finally shoaled, and conditions have stabilised sufficiently to allow the growth of fringing reefs as shown in Stage 5. These and the blanket of later terrigenous sediments are faulted and gently folded in response to continued faulting, thrusting and local collapse in the infrastructure. Where uplift is particularly rapid, folding of the 'cascade' type induced by gravity flow may develop on a regional scale on the flanks of basement swells. This appears to have been the case in Malaita, where the sedimentary sequence and underlying lavas have been extensively folded (Rickwood, 1957; Hackman, 1968b).

This sequence illustrates probable trends in the evolution of eastern Guadalcanal, Malaita and the intervening Indispensable Strait basin.

Petrochemical evolution

The petrochemical trends as illustrated by the variation diagrams (Figures 28–32) show that the igneous rocks of Guadalcanal conform only very broadly with the classic trends described by Kuno, Macdonald and others.

The von Wolff diagram (Figure 30) and the silica/alkalis variation diagram (Figure 28) show a general trend intermediate between the Hawaiian Alkalic and Tholeiitic series (Macdonald and Katsura, 1964). The best comparison is with the 'Solomons Trend' (Figure 29) as described by Stanton (1967), the background to which, in relation to the New Georgia Group, is given by Stanton and Bell (1969).

The petrochemical trends for Guadalcanal not only parallel those for New Georgia but provide complementary

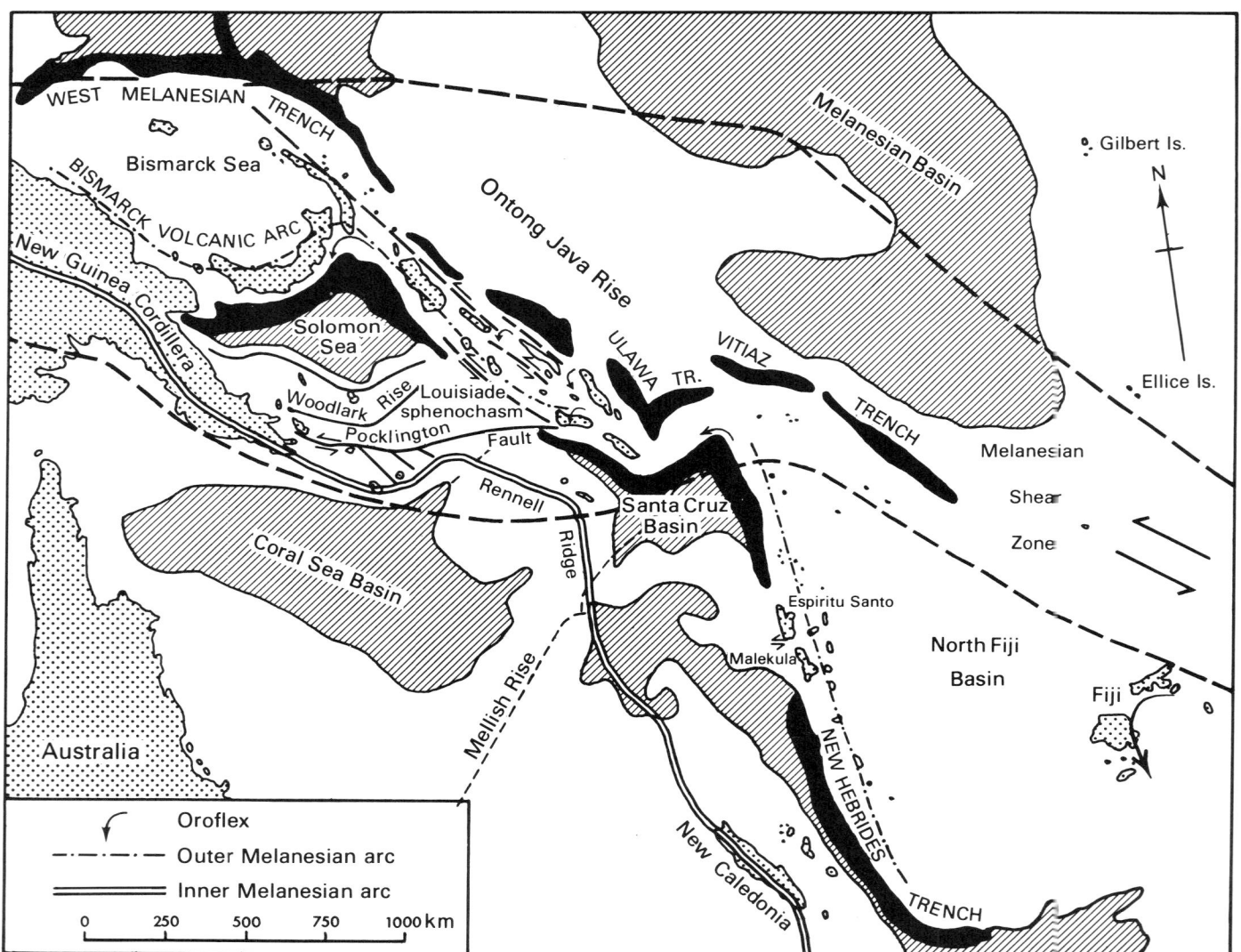

Figure 26 Map of the structural elements of the Melanesian arcs

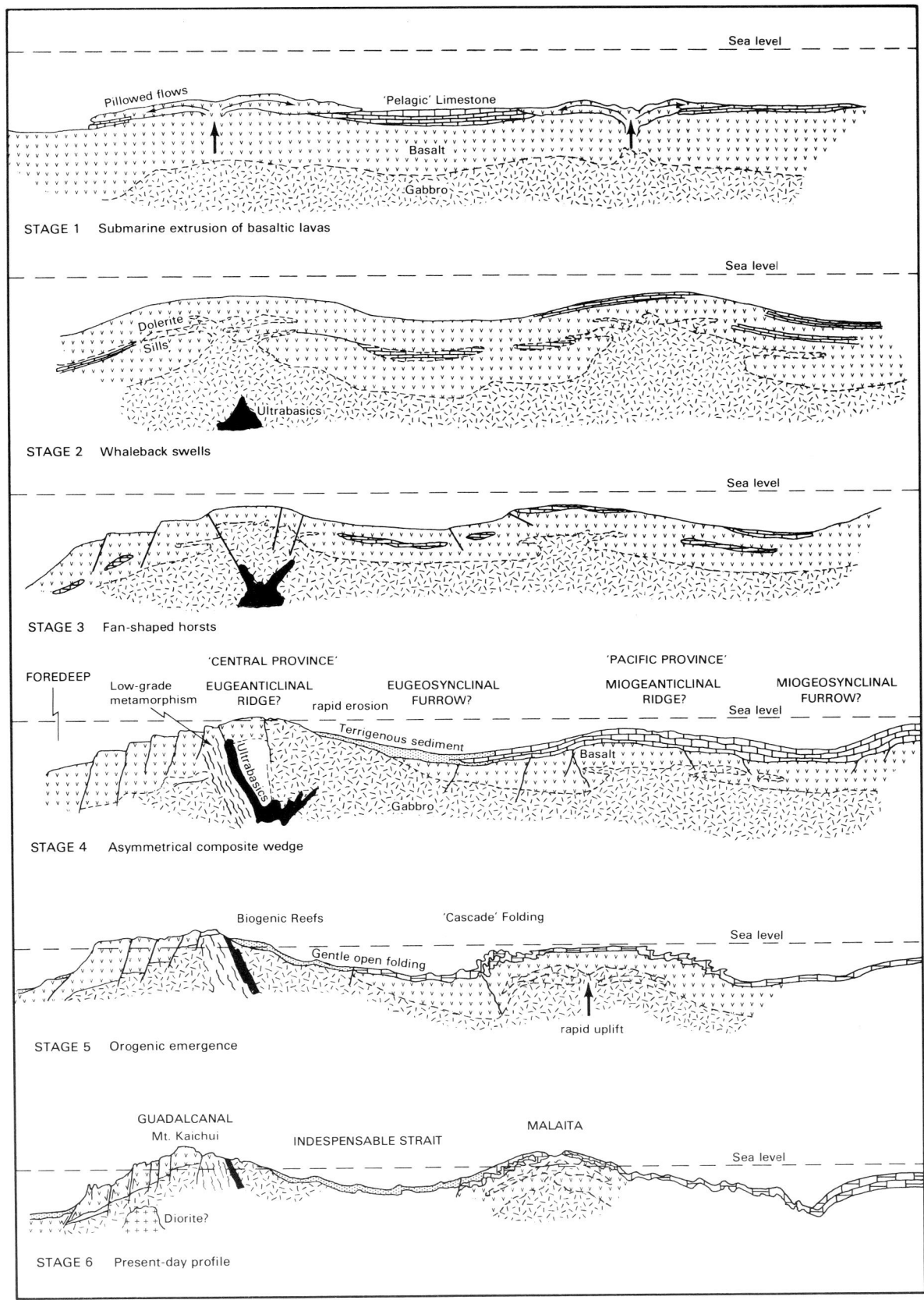

Figure 27 Cross-sections illustrating the evolution of the geanticlinal welts of Guadalcanal and Malaita

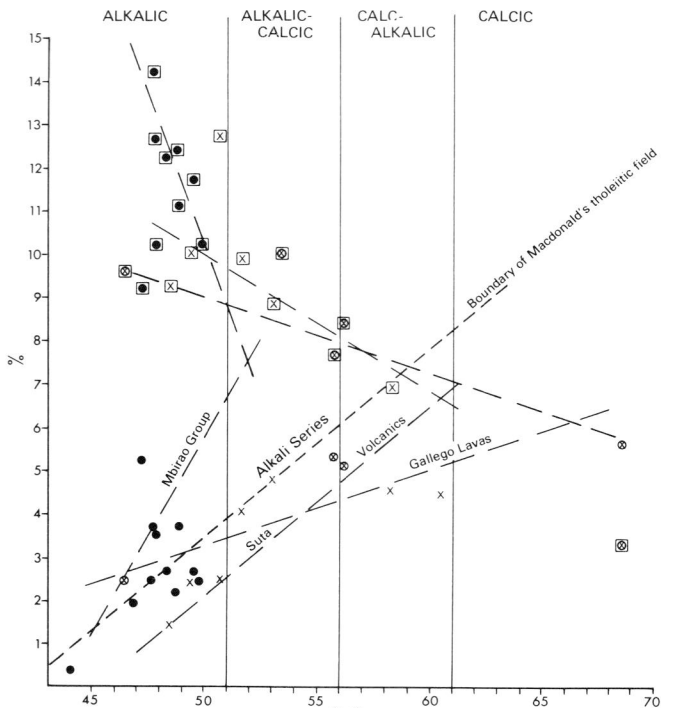

Figure 28 Silica:alkalis variation diagram for igneous rocks of Guadalcanal

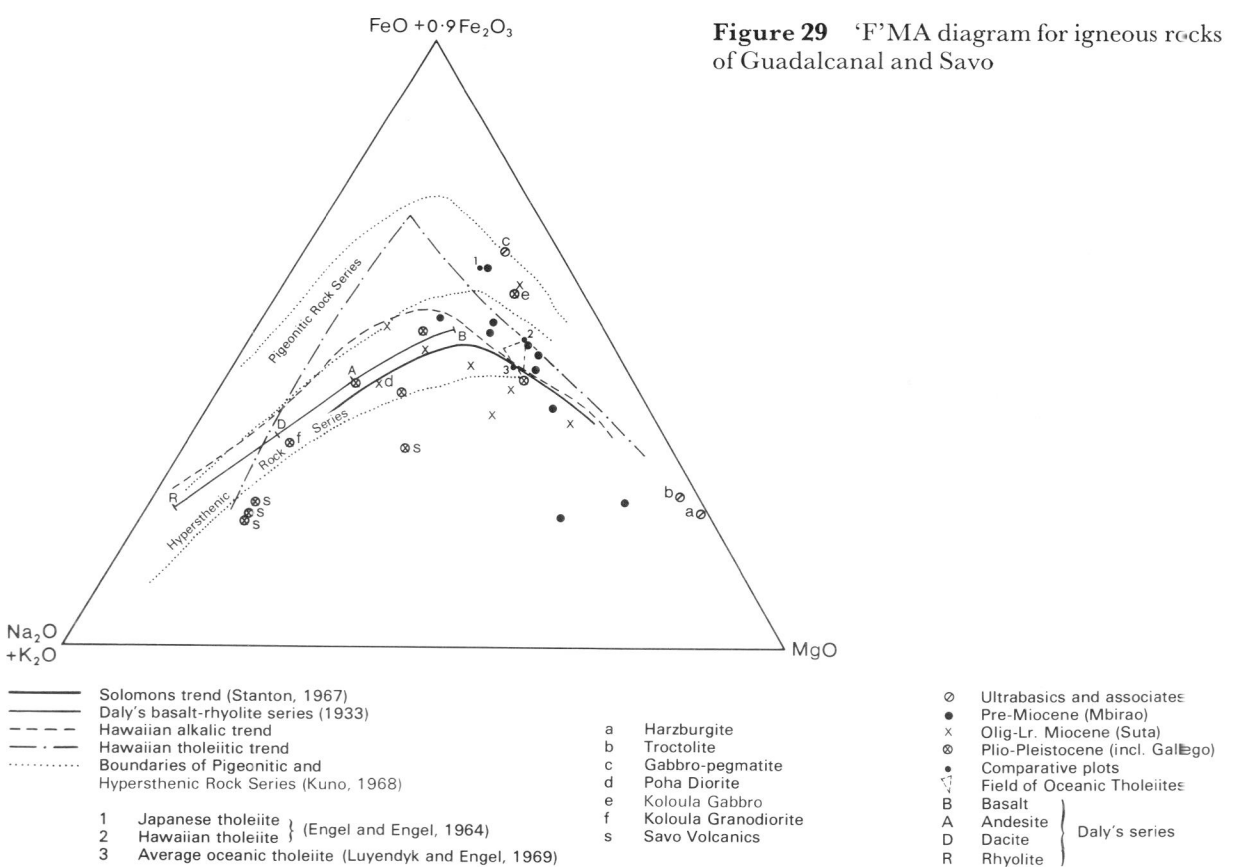

Figure 29 'F'MA diagram for igneous rocks of Guadalcanal and Savo

Figure 30   Von Wolff diagram for igneous rocks of Guadalcanal and Savo

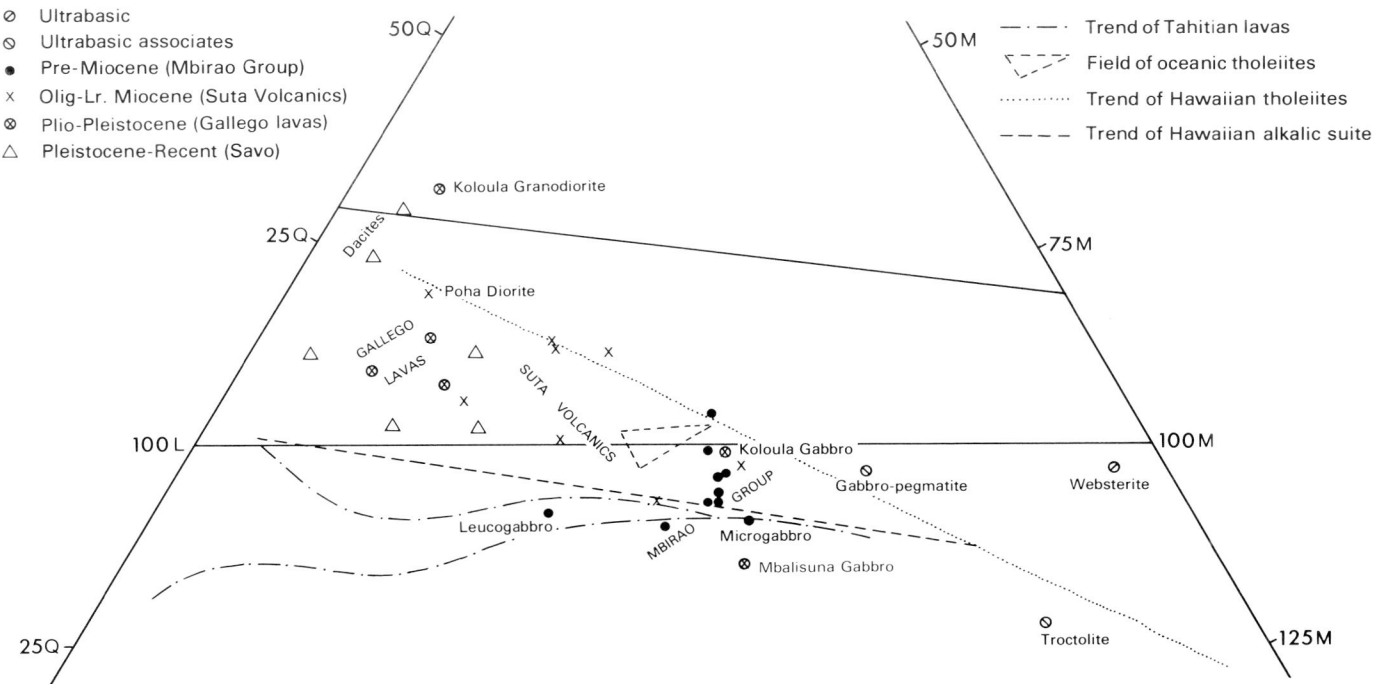

- ⊘ Ultrabasic
- ⊘ Ultrabasic associates
- ● Pre-Miocene (Mbirao Group)
- × Olig-Lr. Miocene (Suta Volcanics)
- ⊗ Plio-Pleistocene (Gallego lavas)
- △ Pleistocene-Recent (Savo)

— · — Trend of Tahitian lavas
⊏⊐ Field of oceanic tholeiites
·········· Trend of Hawaiian tholeiites
— — — Trend of Hawaiian alkalic suite

information: the analyses for the 'basement' rocks of the Mbirao Group, which have no counterpart in New Georgia, fit neatly on the von Wolff diagram in a well defined field, filling a gap between the basaltic andesites and feldsparphyric basalts on the one hand and the picritic basalts on the other (Figure 31, on which all available analyses from the Solomon Islands have been plotted).

The igneous trends appear to be best defined on the von Wolff diagram (modified according to McBirney, 1968) and the 'F' MA diagram, indicating that the significant variables are the relative proportions of mafic and felsic constituents and the degree of iron enrichment (or retention) as compared with weight percent MgO. Since the analyses plot in fields overlapping the boundaries of the 'tholeiite', 'calc-alkali' and such-like series as conceived by different authors, the labelling of trends in such terms is of doubtful value. Coats (1968) employs the ferro-femic index, determined graphically from the 'F' MA diagram, to express the degree of iron enrichment indicated by a particular trend line; this offers a more definitive approach.

The following values for the ferro-femic index show the range of iron enrichment for the Solomons, compared with other well known standards:

East Papuan Field (Jakeš and White, 1969)   38–56
Savo (analyses from Stanton and Jakeš)   45
'Solomons Trend', based on material from
   New Georgia (Stanton, 1967)   63
Calc-alkali trend (Daly, 1933)   65
Suta Volcanics and Gallego Lavas   66
Skaergaard liquids (Wager and Deer, 1939)   95

The occurrence of gabbro in depth throughout the pre-Miocene basement in the Central Province of the Solomons, together with the indications of the existence of a tholeiitic

'root' (derived from the geophysical evidence), suggest that there may be a localised thickening of a gabbroic layer in the upper mantle in the Solomons area.

If the Solomons Arc is to be regarded as strictly autochthonous in the sense that there is no possibility of continental material having been buried, as in eastern Papua, by an overthrust peridotite-gabbro complex (Davies, 1968), the absence of a sialic root points to gabbro or 'tholeiite' as the source material not only for the 'oceanic tholeiites' of the pre-Miocene basement, but also for the basaltic andesites of the Plio-Pleistocene calc-alkaline trend.

In a discussion of the petrogenesis of the New Hebridean volcanic rocks, which exhibit trends comparable to those of the Solomons, Warden (1967) points out that 'the limited evidence available does not support the concept of the existence of sialic rocks in depth'; he considers that andesitic and dacitic lavas could have been derived from tholeiitic basalts by a variable degree of partial melting and differentiation.

Stanton's (1967) hypothesis of derivation of andesitic material from a parent tholeiite involving processes of gas transfer is particularly relevant in this respect, if it is indeed valid to regard the Solomons as a closed system.

Jakeš and White (1969) attempted to correlate the structure of Melanesian arcs with magma types; they considered, on the basis of very limited chemical data, that there was a zonal arrangement of lava types within the New Guinea–New Britain Arc, but that zonation was not distinct within the Solomon Island Arc: rocks from Guadalcanal were regarded as 'calcalkaline'.

However, it is probable that petrochemical zones cannot usefully be delineated within the Solomons area unless carefully tied to detailed structural and stratigraphic studies.

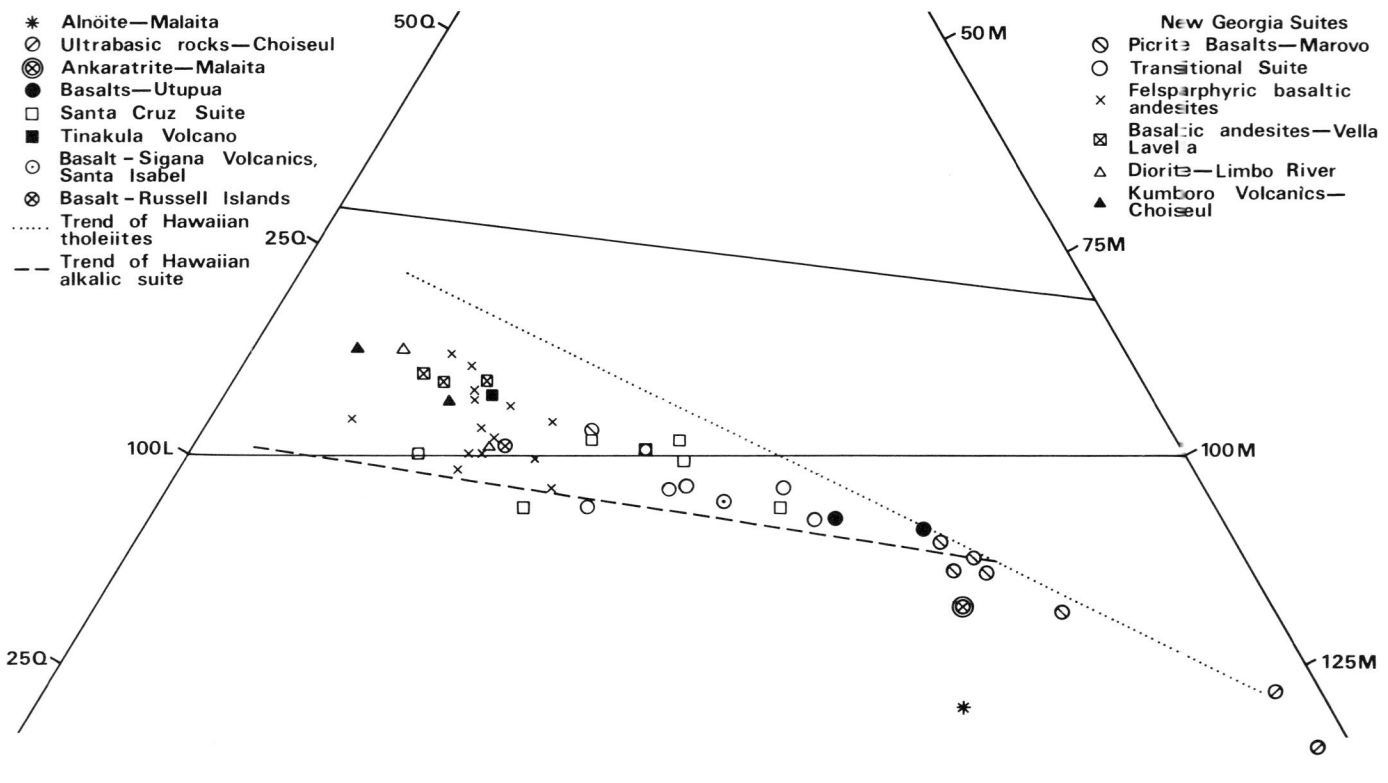

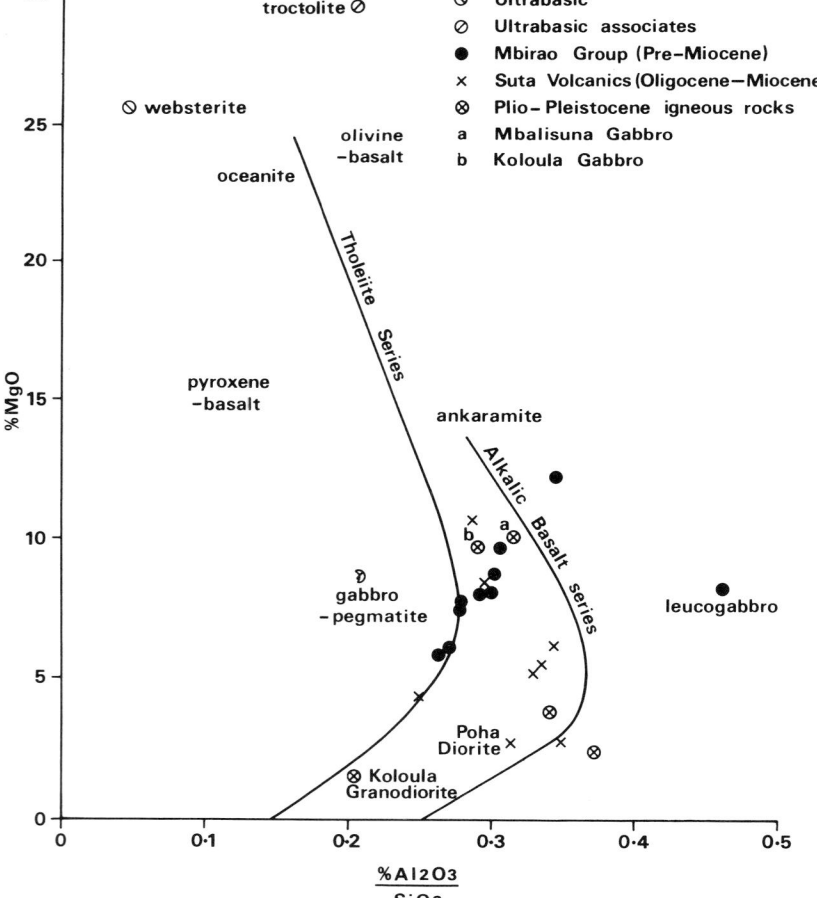

Figure 31 Von Wolff diagram for igneous rocks of the Solomon and Santa Cruz Islands

Figure 32 Murata's variation diagram for igneous rocks of Guadalcanal

Thus Jakeš and White (1969) refer to Cenozoic lavas generally, whereas the data from Guadalcanal indicate that variation in time, and relationship of petrochemical trends to the structural history, is of prime importance. More detailed information may provide a very interesting petrogenetic model for fractured island arcs.

Accordingly it is premature to outline petrochemical zones as Kuno has done (1966) for the Solomons.

In summary, in so far as it is valid to generalise at all on the basis of 27 chemical analyses, the evidence from Guadalcanal indicates that:

1 In the pre-Miocene, the extrusive volcanics were predominantly tholeiitic basalts with oceanic affinities. The more leucocratic gabbros might possibly be regarded as high-level differentiates of a tholeiitic magma, which left a pyroxene-rich ultramafic residuum; the latter was locally injected into 'axial zones' of maximum deformation to form bodies of predominantly harzburgitic composition. Accordingly the problem of the genesis of the ultrabasics must be regarded as bound up with that of the oceanic basalt/gabbro complex, as emphasised by Thayer (1967).

2 Extensive fracturing of the basaltic 'welts' began in the Oligocene–Miocene; a new petrochemical trend was initiated, associated with the eruption of increasing volumes of pyroclastic material, leading from a tholeiitic phase through an intermediate 'basaltic-andesite' phase to an andesitic 'calc-alkaline' trend. The trend culminated in the formation of high-level granodiorite in the Plio-Pleistocene in south Guadalcanal.

At the same time it would appear that gabbroic material, similar to the basalts of the Mbirao Group, was still available high in the crust in the Plio-Pleistocene, for example the Mbalisuna Gabbro and the melanocratic facies of the Koloula Diorite.

The significance of regional stress changes

The complex pattern of metamorphic s-surfaces within the Mbirao Group has been analysed in terms of polyphase deformation (Chapter 3) leading to the conclusion that the derived maximum horizontal regional stress has changed intensity and direction throughout the Cenozoic.

If the derived direction of stress is plotted against time (Figure 33) the graph appears linear, although it must be conceded that considerable uncertainties are attached to some of the age assignments. However, it follows that either the regional stress has changed its orientation progressively in a clockwise direction since the Mesozoic, or it has remained fairly constant in direction and has periodically impressed the marks of a tectonic episode on a block which has rotated consistently in an anticlockwise direction. Clearly a combination of block rotation and change in stress orientation would also have been possible. From Figure 33 the speed of rotation may be assessed at approximately 2° per million years; 70 per cent of the rotation would have been completed by the mid-Miocene, which would accord with the observation that on the geological map sigmoidal bends are manifest primarily in the pre-Miocene basement.

The axial planes of Pliocene folds in Malaita trend north-north-west and north-west (Rickwood, 1957; Hackman, 1968b): the derived axis of maximum stress accordingly

trends east-north-east. Moreover, the disposition of the Kaipito–Korighole Thrust in Santa Isabel (Thompson, 1960; Stanton, 1961) suggests the dominance of stress acting in a north-east to south-west direction. The pre-Miocene basement in these islands has not yet revealed evidence of north-west to south-east stresses comparable to the S_1 phase of the Mbirao Group. However, some evidence from eastern Choiseul (Thompson, 1960) suggests that there may have been comparable stress changes operative in that area; significantly, in eastern Choiseul as in central Guadalcanal, the trend of the ultrabasics is almost normal to the axial trend of the Solomons chain.

According to Hess (1938) 'location of a peridotite belt and dating its intrusion locate the old tectonic axes and date the initiation of the deformation of the zone'. As applied to the Solomons Arc, the 'old tectonic axes' must either have been very sinuous due to primary irregularities or else have been folded about vertical axes since their formation (compare the 'oroclines' of Carey, 1958).

In fact the *en échelon* offsets between the major islands consistently indicate the same sense of movement, that is a sequence of sinistral 'oroflexes' or regional drag folds (Albers, 1967), part of a pattern common to the island arc complexes from Indonesia through to Fiji. On a small scale, each of these bends reflects the configuration of the major re-entrants at either end of the Solomons chain, that is between New Britain and Bougainville at the north-west end and between San Cristobal and Santa Cruz in the south-east.

Figure 33 Variation of azimuth of maximum regional horizontal stress in relation to time

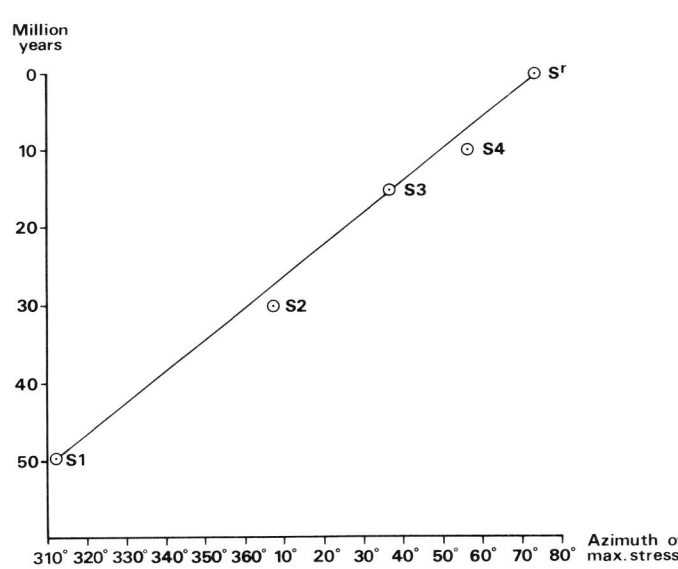

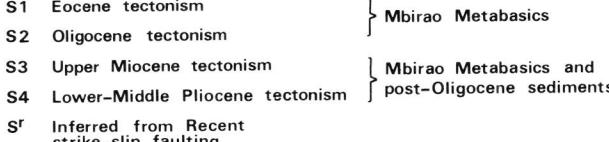

S1   Eocene tectonism   ⎫ Mbirao Metabasics
S2   Oligocene tectonism   ⎬
S3   Upper Miocene tectonism ⎫ Mbirao Metabasics and
S4   Lower–Middle Pliocene tectonism ⎬ post–Oligocene sediments
Sʳ   Inferred from Recent strike–slip faulting

In this connection preliminary palaeomagnetic sampling carried out by Green and Pitt (1967) in New Guinea has given results which accord with the predictions of Carey (1958) that New Guinea would have rotated anticlockwise roughly through a right angle since the end of the Mesozoic.

Limited sampling in the New Hebridean area has not given parallel indications (Tarling, 1966); however, the Melanesian Shear Zone appears to have involved the Solomons–New Guinea area well to the north of the New Hebrides Arc, which has a relatively linear trend, free of sigmoidal bends.

If the Mbirao Block has rotated and is still rotating, Recent seismic activity should be sought along its margins; significantly, perhaps, there is a high density of shallow and intermediate foci between San Cristobal and Guadalcanal (Figure 25). Conceivably the Mbirao Block was originally joined to San Cristobal, also essentially a block of oceanic basalts and intrusive gabbros, but with ultrabasic bodies oriented primarily north-north-west; under the influence of a local component of the Melanesian Shear Zone (?the Pocklington Fault), the north-western end of the San Cristobal block could have become detached and rolled away to the north-west to be welded later onto the Neogene–Quarternary igneous 'vanguard' which now comprises the western half of Guadalcanal.

The possibility that discrete blocks have rotated within orogenic zones elsewhere might shed new light on the local interpretation of polyphase deformation in those areas. Thus Huzita (1969), in a description of the Quaternary tectonics of the Kinki area, southern Japan, attributes anticlockwise rotation of the maximum principal axis of regional stress to changes in direction of upper mantle convection currents: an alternative explanation in terms of clockwise rotation of a block in a steady stress field is perhaps more compatible with the notion of 'sea-floor-spreading' operating at a fairly steady rate and in a constant direction over long periods of time (compare McKenzie and Morgan, 1969). The postulated rotation of blocks in the Solomons is in line with the ideas of van der Linden (1969b), who considered the anticlockwise rotation of a fragmented 'Melanesian complex' with respect to Australia to be superimposed on a north-eastward translation of the entire area, in accord with current interpretations of sea-floor spreading.

Speculations on the Solomons Fractured Arc

Description of the Solomons Arc in terms of classical geosynclinal theory presents some difficulties; the narrow linear orthogeosyncline cannot be identified. The indications are that a pattern of oceanic ridges which in the Late Mesozoic may have approached the linear, has been progressively broken up by tension and sheared into the present-day en échelon configuration. The geophysical evidence is in accord with this view: Rose and others (1968) report that 'Generally inshore [magnetic] anomalies strike parallel to local geological patterns, whereas offshore anomalies strike in a predominantly east–west direction. This east–west strike would substantiate the conclusions of Krause (1966) that the major NW–SE-oriented structure of the Solomon Islands was initiated first, and that only later did it develop into east–west-trending en échelon fractures'. As far as the Solomons are concerned, a quotation from Laubscher (1969) is appropriate: 'if a uniformitarian correlation exists at all, the concept of "geosyncline" must be abandoned: the narrow furrow is to be replaced by a series of ocean basins of considerable variable width.'

However, as a very broad application of the concepts of Aubouin (1965), the Central and Volcanic provinces show primary eugeosynclinal features, whilst the Pacific Province might be regarded as a miogeosynclinal counterpart. The submarine vulcanicity of the pre-Miocene is a characteristic feature of a primary eugeosynclinal zone (Cady, 1950): the orogenic uplift and intermediate to acid magmatic activity of the Plio-Pleistocene are features of Cady's primary mountain-building stage. In the Solomons, however, there is no evidence of an adjacent continental 'craton': the platform of operations appears to have been essentially oceanic.

Small sedimentary basins such as the Mbokokimbo and Lungga–Itina troughs in Guadalcanal may be compared with the taphrogeosynclines of Kay (1951) in that they are fault-bounded basins rather than crustal downbuckles, and have received greywacke-type sediments and tilloid material from sub-aqueous mudstreams. Such fault-bounded basins could be regarded as small-scale versions of the rhombochasms and sphenochasms of Carey (1958), and thus be essentially of tensional origin. Several authors have emphasised the role of tensional break-up or disjunctive spreading in relation to the south-west Pacific; Coleman (1960a), in reference to Choiseul, used the term 'taphrogenic'; Grover (1958a) in an essay on the structure of the central Solomons had used the same term.

Illies (1969) distinguishes taphrogenesis in the intercontinental rift system, due to tension and vertical uplift of sub-crustal swells, from orogenesis, which involves compression and vertical uplift: he cites the circum-Pacific seismic belt as an example of the latter process. However, the evidence from the Solomons does not allow the fractured arc to be categorised unequivocally in this way; certainly the structural pattern analysed in Guadalcanal are complex, but if the deduced dynamic picture is valid, then they are to be interpreted in terms of alternations of two complementary mechanisms:

1 The dominance of north-east to south-west stress, relieved by buckling of the geanticlinal welts accompanied by compression in depth and tensional features along the crests.

2 The relief of stress by deep-seated transcurrent faulting along west-north-west lines, with associated tensional rifting and rotation of discrete blocks of oceanic upper-mantle material.

Coleman (1967) suggested that towards the end of the Mesozoic what is now eastern Australia was bounded by a double arc system which fragmented and 'exploded in tension' to produce the Melanesian Re-entrant, the complex of arcs with abrupt angular junctures between New Britain and Fiji. Differential movement between the arcs could have been accommodated by transcurrent movement, for example along the Charters Towers lineament and its extension along the Mellish Rise to the re-entrant between the Santa Cruz Islands and San Cristobal (Figure 26). These ideas were foreshadowed by Benson (1924): 'the modern volcanic and seismic activity in the New Hebrides is but a continuation of the same controls which produced the Tertiary orogeny and plutonic intrusion in New Caledonia, and successive zones of Permian, Carboniferous, Devonian and early Palaeozoic folding and intrusion which may be traced in southwesterly sequence through the eastern states of Australia'.

The geological history of Guadalcanal is not incompatible with these general lines of thinking. In this context it is interesting to compare Carey's explanation of the origin of Melanesian structures Figure 12A with the stress and faulting patterns illustrated by Figures 12A and 15.

The anomaly of the juxtaposition of oceanic lavas and Upper Cretaceous pelagic oozes of the Pacific Province with the Central Province has been emphasised by Coleman (1966). However, the difficulty disappears if Malaita and the Ontong Java Platform are regarded, not as a miogeosynclinal counterpart to the Central Province, but as parts of an oceanic 'craton' which has been approached from the south by an expanding Outer Melanesian Arc. There is then no equivalent of the Pacific Province in the New Hebrides arc, since the oceanic area to the east has not buckled sufficiently to shoal as in the case of Malaita.

Plate tectonics

Several authors have tried to explain the pattern of island arcs in the south-west Pacific in terms of sub-crustal convection currents. Cullen (1967) envisaged that during the Mesozoic north-east to south-west stress dominated the region and initiated the structural framework of island arcs between Australia and Tonga–Kermadec. Under the influence of a convection stream arising from the Indian–Antarctic Ridge, the crust of the basin as a whole was conceived as having migrated to the north-north-east, to be thrust under the Pacific Basin along the line of volcanic arcs between New Britain and Tonga.

Wright (1966) had appealed to a similar mechanism, but interposed a weak secondary convection cell between the Outer and Inner Melanesian arcs. Van der Linden (1969a) suggested that a sea-floor spreading mechanism, arising from a now extinct or dormant mid-ocean ridge in the south-west Pacific, could explain the structural evolution of the island arcs.

Badgley (1965, p. 421) suggested that 'the north-east-dipping seismic zones in the Solomon and New Hebrides islands might be related to older south-west-flowing currents off the mid-Pacific Mesozoic ridge'. On the same lines Robinson (1968) suggested that the Darwin Rise could have acted as a continental 'foreland' in relation to the Melanesian arcs.

Denham (1969) in a discussion of the earthquake distribution in the Solomons–New Guinea area suggests that the complex pattern is due to the location 'in the zone of interaction between the westward-spreading Pacific "plate" and the northward-moving "plate" containing the Australian continent and the south-east Indian Ocean'. He concluded that although most features might be explained in terms of sea-floor spreading, the low level of seismicity across the central part of the Solomons, associated with the absence of a well-marked trench, cannot so easily be reconciled with the model.

Westwood (1970) also regarded the central Solomons as enigmatic in this respect, suggesting that the seismic zone might be locally vertical with reduced activity, or there might be a series of small blocks related closely to the pattern of surface islands and separated by transform faults. The writer is inclined to favour the latter interpretation, as it accords better with the complex picture which has emerged from structural analysis.

More data are required on dip-slip vectors from within the Solomons Arc; perusal of Figure 26 shows how important components of the Melanesian Shear Zone, such as the Pocklington Fault, transect the Solomons Arc where the regional seismic pattern is broken between Santa Cruz and the western Solomons; possibly the zonal pattern has in fact been smeared out and distorted by deep-seated transcurrent movements. An approach to the understanding of the deep-seated mechanisms must take into account evidence for vertical and horizontal movements as well as the patterns of vulcanicity and seismicity.

The analysis of the Melanesian arcs in terms of Plate Tectonics theory encounters two different types of problem:
1   Interpretations of the seismic and heat-flow patterns in terms of regionally consistent systems.
2   The integration of topographic, geological and geophysical data with such interpretations.

The occurrence of minor seismic belts with shallow focal depth has led different investigators to recognise a number of small subsidiary plates in the east New Guinea–Solomons area:
1   The Solomon Sea Plate (Johnson and Molnar, 1972) or the Solomons Plate (Luyendyk and others, 1973) between the Woodlark Rise and the Planet Trench.
2   The South Bismarck Plate (Johnson and Molnar, 1972) between the mildly seismic North Bismarck Arc and the Bismarck Volcanic Arc.
3   The North Bismarck Plate (Johnson and Molnar, 1972) between the North Bismarck Arc and the West Melanesian Trench.
4   The Bismarck Plate, together with a tentative easterly extension which incorporated the Solomon Islands Arc.

Ripper (1970), interpreting the focal mechanism analyses of large earthquakes for 1963–1967, showed that in the New Britain area the pressure axes of dip-slip earthquakes remain orthogonal to the trench, that is they change trend from north-westerly along the Bismarck Volcanic Arc to north-easterly in the Bougainville area. He was unable to reconcile this with the simple pattern of an Australian Plate underthrusting the Pacific Plate in the Solomons Sea area. However, Curtis (1973a & b) considered this dilemma to be resolved by the recognition of another small plate in the southern Bismarck Sea, the South Bismarck Plate, which was overthrusting the Solomon Sea in an east-south-easterly direction.

Milsom (1970) and Luyendyk and others (1973) interpreted the Woodlark Basin, between the Pocklington Rise and the Woodlark Rise, as a centre of sea-floor spreading; the axis of the basin is regarded as currently active over only about 300 km at the eastern end, that is immediately south-west of the New Georgia group. The Louisiade Sphenochasm concept (Carey, 1958) is thus vindicated in a new guise.

A zone of sea-floor spreading in the Woodlark Basin could explain both the break in the Solomons Trench between Guadalcanal and west New Georgia, and the occurrence of Pliocene–Recent volcanics (the Volcanic Province) in New Georgia and west Guadalcanal, which are associated with taphrogenic zones characterised by swarms of normal faults trending north-north-east.

However, Halunen and von Herzen (1973) interpret the currently active New Georgia volcanic zone as associated with former subduction of the Pacific Plate on the east side of Malaita, 200–250 km to the north-east. They suggest that subduction was blocked by collision between the Solomons Arc and the Ontong Java Plateau, giving rise to a very recent polarity reversal.

The patterns of heat flow and high shear-wave attenuation are anomalous in terms of the classic island arc situation, both showing higher levels on the convex (trench) side of the arc than on the Pacific side: again, these patterns could be considered as related to 'pre-flip' conditions.

Coleman (1975a & b) summarised some of the dilemmas encountered in the interpretation of the Solomons Arc in terms of Plate Tectonics theory. He considers that although the Solomons Arc has been described by Karig and Mammerickx (1972) as having undergone polarity reversal, probably within the last ten million years, 'the case for viewing the Solomons as a NE-facing arc during the Palaeogene is not proven'.

Greater certainty must await determinations of palaeolatitudes, more seismic studies, and detailed information on the sea floor between Australia and the Outer Melanesian Arc. Comparative palaeomagnetic studies of Guadalcanal and Malaita could be very revealing.

REFERENCES

A.B.E.M. 1967. Report on an airborne geophysical survey in the British Solomon Islands. Aktiebolag Elektrisk Malmletning (Stockholm) 1965–1966. Vol. 1, 160 pp., and Vol. 2, 156 pp.

ALBERS, J. P. 1967. Belt of sigmoidal bending and right-lateral faulting in the Western Great Basin. *Bull. Geol. Soc. Am.*, Vol. 78, pp. 143–156.

ALLUM, J. A. E. 1967. *Regional photogeological interpretation of the British Solomon Islands. UN Special Development Program Aerial Geophysical Survey Project Report*. 78 pp. (Honiara, Solomon Islands: Government Printing Office.)

AMHERST (Lord, of Hackney) and THOMSON, B. 1901. The discovery of the Solomons by Alvaro de Mendaña in 1568. *Hakluyt Soc. Publs.*, Series 2, Nos. 7–8, 482 pp.

AMSTUTZ, G. C. 1968. Spilites and spilitic rocks. Pp. 737–753 in *Basalts: The Poldervaart treatise on rocks of basaltic composition*. Vol. 2. H. H. HESS and A. POLDERVAART (Editors). (New York: Wiley.)

ANDERSON, E. M. 1951. *The dynamics of faulting*. 206 pp. (Edinburgh: Oliver and Boyd.)

AUBOUIN, J. 1965. *Geosynclines. Developments in geotectonics*. Vol. 1. 335 pp. (Amsterdam: Elsevier.)

BADGLEY, P. C. 1965. *Structural and tectonic principles*. 521 pp. (New York: Harper and Row, and John Weatherhill.)

BARRUS, R. B. 1975. Final report on Reconnaissance Permit R37, western Guadalcanal. Unpublished report of AMOCO Minerals Solomons Ltd. 20 pp.

— and WEILAND, E. 1975. Final report on Poha Prospect, Guadalcanal, BSIP. Project A-75-31. Unpublished report of AMOCO Minerals Solomons Ltd. 47 pp.

BELOUSSOV, V. V. 1962. *Basic problems in geotectonics*. 809 pp. (New York: McGraw Hill.)

BENSON, W. N. 1924. The structural features of the margin of Australasia. *Trans. Proc. N.Z. Inst.*, Vol. 55, pp. 93–137.

BERSENEV, I. I., BURYANOVA, I. Z., KASIAN, E. D. and LICHT, F. R. 1966. Tuff lavas of southern Sikhote–Alin. Pp. 112–118 in *Tuff lavas and ignimbrites, a survey of Soviet studies*. E. F. COOK (Editor). (Amsterdam: Elsevier.)

BILLINGS, M. P. 1942. *Structural geology*. 473 pp. (New Jersey: Prentice-Hall.)

BLAKE, D. H. 1968. Post-Miocene volcanoes on Bougainville Island, Territory of Papua and New Guinea. *Bull. Volcanol.*, Vol. 32 (1), pp. 121–138.

— and MIEZITIS, Y. 1967. Geology of Bougainville and Buka Islands, New Guinea. *Bull. Bur. Miner. Resour. Geol. Geophys. Aust.*, Vol. 93 (PNG 1). 56 pp.

BUDDINGTON, A. F. 1939. Adirondack igneous rocks and their metamorphism. *Mem. Geol. Soc. Am.*, No. 7. 354 pp.

CADY, W. M. 1950. Classification of geotectonic elements. *Trans. Am. Geophys. Union*, Vol. 31, pp. 780–785.

CAREY, S. W. 1958. Pp. 177–355 in *Continental drift. A symposium*. (Hobart: Geology Dep., Univ. Tasmania.)

— 1963. Syntaphral tectonics and diagenesis. A symposium. 184 pp. (Hobart: Geology Dep., Univ. Tasmania.)

CAROZZI, A. V. 1960. *Microscopic sedimentary petrography*. (New York: Wiley.)

CHAYES, F. 1964. A petrographic distinction between Cenozoic volcanics in and around the open oceans. *J. Geophys. Res.*, Vol. 69, pp. 1573–1578.

— 1965. Titania and alumina content of oceanic and circum-oceanic basalt. *Mineral. Mag.*, Vol. 34, pp. 126–131.

CHINNERY, M. A. 1966. Secondary faulting. *Can. J. Earth Sci.*, Vol. 3, pp. 163–190.

CHIVAS, A. R. 1975. Geochemistry of the Koloula Igneous Complex, Guadalcanal. *Bull. Aust. Soc. Explor. Geophys.*, Vol. 6, pp. 64–65.

— and WILKINS, R. W. T. *In press*. Fluid inclusion studies in relation to hydrothermal alteration and mineralization at the Koloula Porphyry Copper Prospect, Guadalcanal. *Econ. Geol.*

CLOOS, E. 1947. Oolite deformation in South Mountain fold, Maryland. *Bull. Geol. Soc. Am.*, Vol. 58, pp. 843–918.

COATS, R. R. 1968. Basaltic andesites. Pp. 689–736 in *Basalts: The Poldervaart treatise on rocks of basaltic composition*. Vol. 2. H. H. HESS and A. POLDERVAART (Editors). (New York: Wiley.)

COLEMAN, P. J. 1957. Geology of western Guadalcanal. *Colon. Geol. Miner. Resour.*, Vol. 6, No. 3, pp. 288–300.

— 1960a. An introduction to the geology of the island of Choiseul in the Western Solomons, 1957. *Br. Solomon Isl. Geol. Rec.*, (1957–1958), pp. 16–26.

— 1960b. North-central Guadalcanal—An interim geological report. *Br. Solomon Isl. Geol. Rec.*, (1957–1958), pp. 4–13.

— 1963. Tertiary larger foraminifera of the British Solomon Islands, south-west Pacific. *Micropalaeontology*, Vol. 9, pp. 1–38.

— 1965a. Stratigraphical and structural notes on the British Solomon Islands with reference to the First Geological Map, 1962. *Br. Solomon Isl. Geol. Rec.*, Vol. 2 (1959–1962), pp. 17–31.

— 1965b. Tertiary assemblages of larger foraminifera in the Solomon Islands and New Hebrides archipelago. *Contr. Annu. Rep., New Hebrides Geol. Surv.*, pp. 48–51.

— 1966. The Solomon Islands as an island arc. *Nature, London*, Vol. 211, No. 5055, pp. 1249–1251.

— 1967. A possible resolution of the Melanesian re-entrant. Pp. 192–194 in *Upper Mantle Project, 2nd Aust. Prog. Rep., 1965–1967*. (Canberra: Australian Academy of Sciences.)

— 1968. Tsunamis as geological agents. *J. Geol. Soc. Aust.*, Vol. 15, pp. 267–273.

— 1970. Geology of the Solomon and New Hebrides islands, as part of the Melanesian re-entrant, south-west Pacific. *Pac. Sci.*, Vol. 24, No. 3, pp. 289–314.

— 1975a. On island arcs. *Earth Sci. Rev.*, Vol. 11, pp. 47–80.

— 1975b. The Solomons as a non-arc. *Bull. Aust. Soc. Explor. Geophys.* Vol. 6, No. 2/3, pp. 60–61.

and DAY, A. A. 1965. Petroleum possibilities and marked gravity anomalies in north-central Guadalcanal. *Br. Solomon Isl. Geol. Rec.*, Vol. 2 (1959–1962), pp. 112–119.

— and MCTAVISH, R. A. 1964. Association of larger and planktonic foraminifera in single samples from Middle Miocene sediments, Guadalcanal, Solomon Islands, south-west Pacific. *J. R. Soc. W. Aust.*, Vol. 47 (1), pp. 13–24.

— — 1967. Association of early Miocene planktonic and larger foraminifera from the Solomon Islands, south-west Pacific. *Aust. J. Sci.*, Vol. 29, No. 10, p. 373.

COLES, I. G. 1971. Exploration on reconnaissance permit R2, Guadalcanal, BSIP. Unpublished report of Carpentaria Explor. Co. Pty Ltd. 30 pp.

CONNOLLY, H. J. C. 1948. The Sorvohio and Gold Ridge prospects. 6 pp. Unpublished report of Solomon Islands Ministry of Natural Resources.

CONNOLLY, J. W. 1960. The Cape Henslow Iron Formation on Guadalcanal, 1957. *Br. Solomon Isl. Geol. Rec.*, (1957–1958), p. 58.

CONYBEARE, C. E. B. and CROOK, K. A. W. 1968. Manual of sedimentary structures. *Bull. Bur. Miner. Resour. Geol. Geophys. Aust.*, Vol. 102, 327 pp.

COTTON, C. A. 1958. *Geomorphology*. 505 pp. (Christchurch: Whitcombe and Tombs.)

CRAIG, P. 1968. Report on reconnaissance prospecting in the Beaufort Bay to Wanderer Bay area of Guadalcanal. 4 pp. Unpublished report of Solomon Islands Ministry of Natural Resources.

— *In press*. The geology of Santa Cruz, eastern Solomon Islands.

CULLEN, D. J. 1967. Island arc development in the south-west Pacific. *Tectonophysics*, Vol. 4, No. 2, pp. 163–172.

CURTIS, J. W. 1973a. The spatial seismicity of Papua–New Guinea and the Solomon Islands. *J. Geol. Soc. Aust.*, Vol. 20, Part 1, pp. 1–20.

— 1973b. Plate Tectonics and the Papua–New Guinea–Solomon Islands region. *J. Geol. Soc. Aust.*, Vol. 20, Part 1, pp. 21–36.

DALY, R. A. 1933. *Igneous rocks and the depths of the earth.* 598 pp. (New York: McGraw-Hill.)

DAVIES, H. L. 1968. Papuan ultramafic belt. *Rep. 23rd Sess. Int. Geol. Congr., Czechoslovakia, 1968*, Part 1, pp. 209–220.

DE GOLYER and MACNAUGHTON Inc. 1965. Petroleum: Geology and oil prospects of north-central Guadalcanal. *Br. Solomon Isl. Geol. Rec.*, Vol. 2 (1959–1962), pp. 120–123.

DE SITTER, L. U. 1956. *Structural geology*. 552 pp. (New York: McGraw-Hill.)

DENHAM, D. 1969. Distribution of earthquakes in the New Guinea–Solomon Islands region. *J. Geophys. Res.*, Vol. 74, No. 17, pp. 4290–4299.

DEWEY, J. F. 1969. Continental margins: a model for conversion of Atlantic to Andean type. *Earth Planet. Sci. Lett.*, Vol. 6, pp. 189–197.

DIETZ, R. S. 1963. Alpine serpentinites as oceanic rind fragments. *Bull. Geol. Soc. Am.*, Vol. 74, pp. 947–952.

EDWARDS, A. B. 1950. The petrology of the Miocene sediments of the Aure Trough, Papua. *Proc. R. Soc. Victoria*, Vol. 60, pp. 123–148.

EMMONS, R. C. 1969. Strike-slip rupture patterns in sand models. *Tectonophysics*, Vol. 7, No. 1, pp. 71–87.

ENGEL, A. E. J. and ENGEL, C. G. 1964. Igneous rocks of the East Pacific Rise. *Science*, Vol. 146, pp. 477–485.

FAIRBRIDGE, R. W. 1961. The Melanesian border plateau, a zone of crustal shearing in the south-west Pacific. Pp. 137–149 in Structure de la croûte terrestre. *Publ. Bur. Cent. Seismol. Int.*, Vol. A22.

FISHER, R. V. 1960. Classification of volcanic breccias. *Bull. Geol. Soc. Am.*, Vol. 71, pp. 973–979.

FOLK, R. L. 1959. Practical petrographic classification of limestones. *Bull. Am. Ass. Petrol. Geol.*, Vol. 43, pp. 1–38.

— 1962. Spectral subdivision of limestone types. *Symp. Am. Assoc. Pet. Geol. Mem.*, No. 1, pp. 62–84.

GARDNER, M. E. 1972a. Annual report on special prospecting licence C83, Mbetilonga, Guadalcanal, BSIP. Utah Development Co., Exploration Dep. Rep., No. 182, 24 pp. [Unpublished].

— 1972b. Final report on special prospecting licence C89, Avuavu. 8 pp. Unpublished Report of Utah Development Co. Exploration Dep., Aug. 1972.

— 1973. Annual report for 1972, Mbetilonga Prospect, BSIP. Utah Development Co., Exploration Dep Rep., No. 210. 23 pp. [Unpublished].

GASS, I. G. 1968. Is the Troodos massif of Cyprus a fragment of Mesozoic ocean floor? *Nature, London*, Vol. 220, pp. 39–42.

GLAESSNER, M. F. 1950. Geotectonic position of New Guinea. *Bull. Am. Assoc. Pet. Geol.*, Vol. 34, pp. 856–881.

GRASSO, V. G. 1968. The TiO_2 frequency in volcanic rocks. *Geol. Rundsch.*, Vol. 57, pp. 930–935.

GREEN, D. H. 1969. The origin of basaltic and nephelinitic magmas in the earth's mantle. *Tectonophysics*, Vol. 7, Nos. 5–6, pp. 409–422.

— and PITT, R. P. B. 1967. Suggested rotation of New Guinea. *J. Geomagn. Geoelectr.*, Vol. 19, No. 4, pp. 317–321.

GROVER, J. C. 1955a. Betilonga Basin area, Guadalcanal. Pp. 82–86 *in* Geology, mineral deposits and prospects of mining development in the British Solomon Islands Protectorate. *Interim Mem. Geol. Surv. Br. Solomon Isl.*, No. 1. 108 pp.

— 1955b. Guadalcanal. Pp. 29–34 *in* Geology, mineral deposits and prospects of mining development in the British Solomon Islands Protectorate. *Interim Mem. Geol. Surv. Br. Solomon Isl.*, No. 1. 108 pp.

— 1955c. The history of exploration and mining ventures in the Solomons. Pp. 8–15 *in* Geology, mineral deposits and prospects of mining development in the British Solomon Islands Protectorate. *Interim Mem. Geol. Surv. Br. Solomon Isl.*, No. 1. 108 pp.

— 1955d. Gold Ridge field, Guadalcanal. Chapter 21, pp. 65–77 *in* Geology, mineral deposits and prospects of mining development in the British Solomon Islands Protectorate. *Interim Mem. Geol. Surv. Br. Solomon Isl.*, No. 1. 108 pp.

— 1955e. Lignite deposits, S.W. Guadalcanal. Chapter 28, pp. 97–98 *in* Geology, mineral deposits and prospects of mining development in the British Solomon Islands Protectorate. *Interim Mem. Geol. Surv. Br. Solomon Isl.*, No. 1. 108 pp.

— 1955f. The prospects of oil accumulations. Chapter 32, p. 103 *in* Geology, mineral deposits and prospects of mining development in the British Solomon Islands Protectorate. *Interim Mem. Geol. Surv. Br. Solomon Isl.*, No. 1. 108 pp.

— 1955g. The Sorvohio Valley alluvial flats, Guadalcanal. Chapter 22, pp. 78–79 *in* Geology, mineral deposits and prospects of mining development in the British Solomon Islands Protectorate. *Interim Mem. Geol. Surv. Br. Solomon Isl.*, No. 1. 108 pp.

— 1955h. The Suta Goldfields, Guadalcanal. Chapter 23, pp. 81–82 *in* Geology, mineral deposits and prospects of mining development in the British Solomon Islands Protectorate. *Interim Mem. Geol. Surv. Br. Solomon Isl.*, No. 1. 108 pp.

— 1958a. A regional picture of the central Solomons. Pp. 31–39 *in* The Solomon Islands—geological exploration and research, 1953–1956. *Mem. Geol. Surv. Br. Solomon Isl.*, No. 2. 151 pp.

— 1958b. Gold Ridge, Guadalcanal—discovery of gold-bearing bodies, implications of sample assays, and future prospects. Pp. 63–80 *in* The Solomon Islands—geological exploration and research, 1953–1956. *Mem. Geol. Surv. Br. Solomon Isl.*, No. 2. 151 pp.

— 1958c. The pyritic deposit in the Tina River headwaters. P. 81 *in* The Solomon Islands—geological exploration and research, 1953–1956. *Mem. Geol. Surv. Br. Solomon Isl.*, No. 2. 151 pp.

— 1958d. Note on gold in the Vahanambo River, Guadalcanal. P. 82 *in* The Solomon Islands—geological exploration and research, 1953–1956. *Mem. Geol. Surv. Br. Solomon Isl.*, No. 2. 151 pp.

— 1958e. The Redstone Ridge magnetite deposits, western Guadalcanal. Pp. 60–63 *in* The Solomon Islands—geological exploration and research, 1953–1956. *Mem. Geol. Surv. Br. Solomon Isl.*, No. 2. 151 pp.

— 1960. The beach sands of Guadalcanal—notes on J. W. Connolly's investigations. *Br. Solomon Isl. Geol. Rec.*, 1957–1958, pp. 54–56.

— 1965. Notes on a first map of mineral occurrences in the Solomon Islands, 1962. *Br. Solomon Isl. Geol. Rec.*, Vol. 2 (1959–1962), pp. 48–71.

— 1967. Forecasting of earthquakes. *Nature, London*, Vol. 213, pp. 686–687.

— 1968a. Record of gravity, magnetic, bathymetric, aerial and crustal surveys in the British Solomon Islands, 1963–1966. *Br. Solomon Isl. Geol. Rec.*, Vol. 3 (1963–1967), pp. 110–116.

— 1968b. The British Solomon Islands: some geological implications of the gravity data, 1966 *in* The crust and upper mantle of the Pacific area. *Monogr. Am. Geophys. Union*, Vol. 12, pp. 296–306.

— 1968c. Notes on diamond drilling at Hidden Valley, N.W. Guadalcanal, 1963. *Br. Solomon Isl. Geol. Rec.*, Vol. 3 (1963–1967), pp. 75–77.

— and PUDSEY-DAWSON, P. A. 1958. Notes on the geology of eastern Guadalcanal. Pp. 56–60 *in* The Solomon Islands—geological exploration and research, 1953–1956. *Mem. Geol. Surv. Br. Solomon Isl.*, No. 2. 151 pp.

GUPPY, H. B. 1887. The Solomon Islands: their geology, general features and suitability for colonisation. 152 pp. (London: Swan Sonnenschein & Co.)

HACKMAN, B. D. 1968a. *A guide to the spelling and pronunciation of place names in the British Solomon Islands Protectorate.* 18 pp. (Honiara, Solomon Islands: Government Printing Office.)

— 1968b. Observations on folding in the Oligocene–Miocene limestones of central Kwara'ae, Malaita. *Br. Solomon Isl. Geol. Rec.*, Vol. 3 (1963–1967), pp. 47–50.

— 1968c. The geology of east and central Guadalcanal: a preliminary statement, 1966. *Br. Solomon Isl. Geol. Rec.*, Vol. 3, pp. 16–25.

— 1973. The Solomon Islands fractured arc. Pp. 179–191 in *The Western Pacific: Island arcs, marginal seas, geochemistry.* P. J. COLEMAN, (Editor). (Perth: University of Western Australia Press.)

— 1977. The geology of the Honiara area, Guadalcanal. *Bull. Geol. Surv. Solomon Isl*, No. 3.

— *In preparation.* The geology of the Tiaro Bay area, Guadalcanal. *Bull. Geol. Surv. Solomon Isl*

HALUNEN, A. J., Jr and HERZEN, R. P. von 1973. Heat flow in the western equatorial Pacific Ocean. *J. Geophys. Res.*, Vol. 78, pp. 5195–5208.

HAMMOND, B. and GOSS, B. 1972. Reconnaissance geological survey of some beaches in the British Solomon Islands Proctectorate. 3 pp. Unpublished report of R. Hare & Associates Pty Ltd, Melbourne.

HATCH, F. H., WELLS, A. K. and WELLS, M. K. 1949. *The petrology of the igneous rocks.* 515 pp. (London: Thomas Murby.)

HESS, H. H. 1938. A primary peridotite magma. *Am. J. Sci.*, Vol. 35, pp. 321–344.

HILL, J. H. 1960. Further exploration of the Betilonga area of Guadalcanal, 1957. *Br. Solomon Isl. Geol. Rec.*, (1957–1958), pp. 81–94.

HILLS, E. S. 1963. 2nd Ed. *Elements of structural geology.* 502 pp. (Chapman and Hall Ltd & Science Paperbacks.)

HODGSON, J. H. 1958. Direction of displacement in western Pacific earthquakes. Pp. 69–86 *in* Contributions in Geophysics in honour of Beno Gutenberg, Vol. 1.

HUGHES, G. W. *In preparation.* The geology of the Lungga Basin, Guadalcanal. *Bull. Geol. Surv. Solomon Isl.*

HUZITA, K. 1969. Tectonic development of south-west Japan in the Quaternary Period. *J. Geosci. Osaka City Univ.*, Vol. 12, Part 5, pp. 53–70.

ILLIES, J. H. 1969. An intercontinental belt of the world rift system. *Tectonophysics*, Vol. 8, No. 1, pp. 5–29.

ISACKS, B. L., OLIVER, J. and SYKES, L. R. 1968. Seismology and the new global tectonics. *J. Geophys. Res.*, Vol. 73, pp. 5855–5899.

— — — 1969. Focal mechanisms of deep and shallow earthquakes in the Tonga–Kermadec region and the tectonics of island arcs. *Bull. Geol. Soc. Am.*, Vol. 80, pp. 1443–1470.

JAKEŠ, P. and WHITE, A. J. R. 1969. Structure of the Melanesian arcs and correlation with distribution of magma types. *Tectonophysics*, Vol. 8, pp. 223–236.

JARVIS, D. M. 1970. Report on a geochemical stream-sediment survey. Reconnaissance permit R5, S.E. Guadalcanal. 9 pp. Utah Development Co., Exploration Dep. Rep. for Dec. 1970. [Unpublished].

JOHANNSEN, A. 1939. A descriptive petrography of the igneous rocks. 4 volumes. 318, 428, 360 and 523 pp. (Chicago: University of Chicago Press.)

JOHNSON, T. and MOLNAR, P. 1972. Focal mechanisms and plate tectonics of the south-west Pacific. *J. Geophys. Res.*, Vol. 77, No. 26, pp. 5000–5032.

JONES, O. A. 1949. Report on Gold Ridge, Guadalcanal, British Solomon Islands Protectorate. 8 pp. Unpublished report of Solomon Islands Ministry of Natural Resources.

JOPLIN, G. A. 1968. *A petrography of Australian igneous rocks.* 253 pp. (Sydney: Angus and Robertson.)

KARIG, D. D. and MAMMERICKX, J. 1972. Tectonic framework of the New Hebrides island arc. *Mar. Geol.*, Vol. 12, pp. 187–205.

KAY, M. 1951. North American geosynclines. *Mem. Geol. Soc. Am.*, No. 48. 143 pp.

KNILL, J. L. 1960. A classification of cleavages, with special reference to the Craignish district of the Scottish Highlands. *Rep. 21st Sess. Int. Geol. Congr., Copenhagen*, 1960, Part 18, pp. 317–325.

KNOPF, E. B. 1931. Retrogressive metamorphism and phyllonitization. *Am. J. Sci.*, Vol. 21, No. 5, pp. 1–27.

KRAUSE, D. C. 1962. Bathymetry and geologic structure of the North Tasman Sea–Coral Sea–South Solomon Sea region of the south-western Pacific Ocean. *Mem. N.Z. Oceanogr. Inst.*, No. 41. 48 pp.

— 1965. Equatorial shear zone in the world rift system. Report of the Symposium on the International Upper Mantle Project. *Geol. Surv. Can. Pap.*, No. 66, p. 14.

KRUMBEIN, W. C. and SLOSS, L. L. 1963. Stratigraphy and sedimentation. 660 pp. (San Francisco: W. H. Freeman & Co.)

KUNO, H. 1959. Origin of Cenozoic petrographic provinces of Japan and surrounding area. *Bull. Volcanol.*, Ser. 2, Vol. 20, pp. 37–76.

— 1960. High-alumina basalt. *J. Petrol.*, Vol. 1, No. 2, pp. 121–145.

— 1966. Lateral variation of basalt magma type across continental margins and island arcs. *Bull. Volcanol.*, Vol. 29, pp. 195–222.

— 1968. Differentiation of basalt magmas. Pp. 623–688 in *Basalts: The Poldervaart Treatise on Rocks of Basaltic Composition.* Vol. 2. H. H. HESS and A. POLDERVAART (Editors). (New York: Wiley.)

LAUBSCHER, H. 1969. Mountain building. *Tectonophysics*, Vol. 7, Nos. 5–6, pp. 551–563.

LAUDON, T. S. 1968. Land gravity survey of the Solomon and Bismarck Islands. Pp. 279–295 in The crust and upper mantle of the Pacific area. *Monogr. Am. Geophys. Union*, No. 12.

LE PICHON, X. 1968. Sea-floor spreading and continental drift. *J. Geophys. Res.*, Vol. 73, p. 3661.

LINDEN, W. J. M. VAN DER. 1969a. Extinct mid-ocean ridges in the Tasman Sea and in the western Pacific. *Earth Planet. Sci. Lett.*, Vol. 6, pp. 483–490.

— 1969b. Rotation of the Melanesian complex and of west Antarctica—a key to the configuration of Gondwana? *Palaeogeogr. Palaeoclimatol. Palaeoecol.*, Vol. 6, No. 1, pp. 37–44.

LI, SZE-KUANG. 1964. Vortex structure and other problems relating to the development of geotectonic systems of north-western China. *Int. Geol. Rev.*, Vol. 6, No. 6, pp. 953–978; No. 7, pp. 1177–1216.

LUYENDYK, B. P. and ENGEL, C. G. 1969. Petrological, magnetic and chemical properties of basalt dredged from an abyssal hill in the north-east Pacific. *Nature, London*, Vol. 223, No. 5210, pp. 1049–1050.

— MACDONALD, K. C. and BRYAN, W. B. 1973. Rifting history of the Woodlark Basin in the southwest Pacific. *Bull. Geol. Soc. Am.*, Vol. 84, pp. 1125–1134.

MCBIRNEY, A. R. 1968. Genetic relations of volcanic rocks of the Pacific Ocean. *Geol. Rundsch.*, Vol. 57, pp. 21–33.

MACDONALD, G. A. 1960. Dissimilarity of Continental and Oceanic rock types. *J. Petrol.*, Vol. 1, No. 2, pp. 172–177.

— and KATSURA, T. 1964. Chemical composition of Hawaiian lavas. *J. Petrol.*, Vol. 5, No. 1, pp. 82–133.

MCKEE, E. D. 1938. Original structures in Colorado flood deposits of Grand Canyon. *J. Sediment. Petrol.*, Vol. 8, pp. 77–83.

— CHRONIE, J. and LEOPOLD, E. B. 1959. Sedimentary belts in lagoon of Kapingamarangi atoll. *Bull. Am. Assoc. Pet. Geol.*, Vol. 43, pp. 501–562.

— and WEIR, G. W. 1953. Terminology for stratification and cross-stratification in sedimentary rocks. *Bull. Geol. Soc. Am.*, Vol. 64, pp. 381–389.

MACKENZIE, D. H. 1968a. Geochemical drainage sampling in the mountains behind Talise, southern Guadalcanal, 1965. *Br. Solomon Isl. Geol. Rec.*, Vol. 3 (1963–1967), pp. 88–93.

— 1968b. Geochemical drainage sampling of the Mbetikama and Mbetilonga areas, Guadalcanal, 1965. *Br. Solomon Isl. Geol. Rec.*, Vol. 3 (1963–1967), pp. 81–87.

MCKENZIE, D. P. and MORGAN, W. J. 1969. Evolution of triple junctions. *Nature, London*, Vol. 224, pp. 125–133.

MCKINSTRY, H. E. 1953. Shears of the second order. *Am. J. Sci.*, Vol. 251, pp. 401–414.

MCTAVISH, R. A. 1968. Planktonic foraminifera from north-central Guadalcanal. *Br. Solomon Isl. Geol. Rec.*, Vol. 3 (1963–1967), pp. 58–65.

MALAKHOV, I. A. 1964. Some problems in the nomenclature of Uralian ultrabasic rocks. *Int. Geol. Rev.*, Vol. 6, pp. 1413–1415.

MALLICK, D. I. J. 1969. Pentecost. *Annu. Rep. New Hebrides Geol. Surv. [for] 1967*, pp. 20–25.

MARANZANA, F. 1968. *Eleven exploration prospects of minor interest in the B.S.I.P. UN Special Development Program, Aerial Geophysical Survey Project Report.* 31 pp. (Honiara, Solomon Islands: Government Printing Office.)

MARSHALL, C. E., COLEMAN, P. J., STANTON, R. L., LEVETT, B. M. and RICKWOOD, F. K. 1957. Geological reconnaissance of parts of the central islands of the British Solomon Islands Protectorate. *Colon. Geol. Miner. Resour.*, Vol. 6, No. 3, pp. 267–306.

MASASHIGE, H. 1969. Joint system of the Rokko Mountain Range. *J. Geosci. Osaka City Univ.*, Vol. 12, No. 3, p. 25.

MASON, D. R. 1975. Geochemistry of intrusive rock suites and related porphyry copper mineralisation in the Papua-New Guinea–Solomon Islands region. 241 pp. Unpublished PhD thesis, Australian National University.

MENARD, H. W. 1964. *Marine geology of the Pacific.* 271 pp. (New York: McGraw–Hill.)

MILSOM, J. S. 1970. Woodlark Basin, a minor centre of sea-floor spreading in Melanesia. *J. Geophys. Res.*, Vol. 73, No. 18, pp. 5925–5941.

MITCHELL, A. H. G. 1969. Raised reef-capped terraces and Plio-Pleistocene sea level changes, north Malekula, New Hebrides. *J. Geol.*, Vol. 77, No. 1, pp. 56–67.

— and WARDEN, A. J. 1971. Geological evolution of the New Hebrides island arc. *J. Geol. Soc. Lond.*, Vol. 127, pp. 501–529.

MOODY, J. D. and HILL, M. J. 1956. Wrench-fault tectonics. *Bull. Geol. Soc. Am.*, Vol. 67, pp. 1207–1248.

MURATA, K. J. 1960. A new method of plotting chemical analyses of basaltic rocks. *Am. J. Sci.*, Vol. 258A, pp. 247–252.

NEWMAN, D. 1968. Ghunukeo copper prospect, S.W. Guadalcanal. *Br. Solomon Isl. Geol. Rec.*, Vol. 3 (1963–1967), pp. 106–108.

OFFICER, C. B. 1955. South-west Pacific crustal structures. *Trans. Am. Geophys. Union*, Vol. 36, pp. 449–459.

O'SULLIVAN, K. N., ENGLISH, P. W., THOMAS, G., KEMP, G., and REBEK, R. J. 1975. Final report on reconnaissance permit R34, Chimiu (S.E. Guadalcanal), British Solomon Islands Protectorate. 75 pp. Unpublished report of CRA Exploration Pty Ltd.

PAGE, R. W. and McDOUGALL, I. 1970. Potassium–argon dating of the Tertiary f1–2 stage in New Guinea and its bearing on the geological time-scale. *Am. J. Sci.*, Vol. 269, pp. 321–342.

PEACOCK, M. A. 1931. Classification of igneous rock series. *J. Geol.*, Vol. 39, pp. 54–67.

PETTIJOHN, F. J. 1957. Sedimentary rocks. 718 pp. (New York: Harper and Bros.)

PISARSKI, J. P. 1968. *Regional geochemical reconnaissance of beach sediments from the mouths of principal rivers in the BSIP. Aerial Geophysical Survey Project Report, UNDP, British Solomon Islands Protectorate, 1965–1968.* 50 pp. (Honiara: Government Printer.)

PLUMLEY, W. J., RISLEY, G. A., GRAVES, R. W., Jr. and KALEY, M. E. 1962. Energy index for limestone interpretation and classification. *Symp. Am. Assoc. Pet. Geol.*, Mem. 1, pp. 85–107.

POLDERVAART, A. 1964. Chemical definition of alkali basalts and tholeiites. *Bull. Geol. Soc. Am.*, Vol. 75, pp. 229–232.

PRINZ, M. 1967. Geochemistry of basaltic rocks. Trace elements. Pp. 271–323 in *Basalts: The Poldervaart treatise on rocks of basaltic composition.* Vol. 1. H. H. HESS and A. POLDERVAART (Editors). (New York: Wiley.)

PUDSEY-DAWSON, P. A. and THOMPSON, R. B. M. 1958. The detailed geological survey of western Guadalcanal, 1954. Pp. 43–56 in The Solomon Islands—Geological exploration and research, 1953–1956. *Mem. Geol. Surv. Br. Solomon Isl.*, No. 2. 151 pp.

PURVIS, J. G. 1975. Report on special prospecting licence C115, Gold Ridge, Guadalcanal, BSIP. 86 pp. Unpublished report of CRA Exploration Pty Ltd.

RAMSAY, J. G. 1967. *Folding and fracturing of rocks.* 568 pp. (New York: McGraw–Hill.)

REBEK, R. J. and THOMAS, G. 1973. Final report on bauxite investigations, R29—Mt Roundhead (Vaturanga–Nggae), BSIP. 7 pp. Unpublished report of CRA Exploration Pty Ltd.

REVELLE, R. R., BRAMLETTE, M., ARRHENIUS, G. and GOLDBERG, E. D. 1955. Pelagic sediments of the Pacific. *Spec. Pap. Geol. Soc. Am.*, Vol. 62, pp. 221–235.

RICHARDS, J. R., COOPER, J. A., WEBB, A. W. and COLEMAN, P. J. 1966. Potassium-argon measurements of the age of basal schists in the British Solomon Islands. *Nature, London*, Vol. 211, No. 5055, pp. 1251–1252.

RICKWOOD, F. K. 1957. Geology of the island of Malaita. *Colon. Geol. Miner. Resour.*, Vol. 6, No. 3, pp. 300–306.

RIPPER, I. D. 1970. Global tectonics and the New Guinea–Solomon Islands region. *Search*, Vol. 1, Part 5, pp. 226–232.

ROBINSON, G. P. 1967. Geomorphology and structure morphotectonics of the Cumberland Peninsula, Santo. *New Hebrides Geol. Surv. Contrib. Annu. Rep. 1965*, pp. 40–46.

— 1968. *Biostratigraphy and structure of Espiritu Santo, New Hebrides archipelago, south-west Pacific.* 317 pp. Unpublished thesis presented for degree of Doctor of Philosophy at University of Western Australia.

— 1969. The geology of North Santo. *New Hebrides Geol. Surv. Reg. Rep.* 80 pp.

RODDA, P. and BAND, R. B. 1967. Geology of Viti Levu. *N.Z. J. Geol. Geophys.*, Vol. 10, No. 5, pp. 1179–1180.

ROSE, J. C., WOOLLARD, G. P. and MALAHOFF, A. 1968. Marine gravity and magnetic studies of the Solomon Islands. Pp. 379–410 in The crust and upper mantle of the Pacific area. *Monogr. Am. Geophys. Union*, 12.

SCHMITT, H. A. 1966. The porphyry copper deposits in their regional setting. Pp. 17–33 in *Geology of the porphyry copper deposits.* S. R. TITLEY and C. L. HICKS (Editors). (University of Arizona Press.)

SHARFMAN, V. S. 1968. Average chemical properties of spilites. *Dokl. Akad. Nauk SSSR.*, Vol. 180, No. 1, pp. 202–203.

SHARPE, C. F. S. 1938. *Landslides and related phenomena.* 137 pp. (Columbia University Press.)

SHIDÔ, F. and MIYASHIRO, A. 1959. Hornblendes of basic metamorphic rocks. *J. Fac. Sci. Univ. Tokyo*, Series 2, Vol. 12, p. 85.

SMITH, B. 1961. Ball or pillow-form structures in sandstones. *Geol. Mag.*, Vol. 53, pp. 146–156.

SNELLING, N. J., INGRAM, I. H. and CHAN, K. P. 1970. K:Ar age determinations on samples from the British Solomon Islands Protectorate. *Inst. Geol. Sci. Geochem. Div. Isotope Geol. Unit Rep.*, No. 70–14.

STANTON, R. L. 1961. Explanatory notes to accompany a first geological map Santa Ysabel, British Solomon Islands Protectorate. *Overseas Geol. & Miner. Resour.*, Vol. 8, No. 2, pp. 127–149.

— 1965. Gold: Alluvial deposits of the Sorvohio River, Guadalcanal. *Br. Solomon Isl. Geol. Rec.*, Vol. 2 (1959–1962), pp. 107–113.

— 1967. A numerical approach to the andesite problem. *Proc. K. Ned. Akad. Wet., Amsterdam*, Series B70, Part 2, pp. 176–216.

— and BELL, J. D. 1969. Volcanic and associated rocks of the New Georgia group, British Solomon Islands Protectorate. *Overseas Geol. & Miner. Resour.*, Vol. 10, No. 2, pp. 113–145.

STODDART, D. R. 1969a. Geomorphology of the Solomon Islands coral reefs. *Philos. Trans. R. Soc. London*, Series B, Vol. 255, pp. 355–382.

— 1969b. Sand cays of eastern Guadalcanal. *Philos. Trans. R. Soc. London*, Series B, Vol. 255, pp. 403–432.

STRINGHAM, B. 1966. Igneous rock types and host rocks associated with porphyry copper deposits. Pp. 35–40 in *Geology of the porphyry copper deposits.* S. R. TITLEY and C. L. HICKS (Editors). (University of Arizona Press.)

TARLING, D. H. 1966. Results of a palaeomagnetic reconnaissance of the New Hebrides and New Caledonia. *Tectonophysics*, Vol. 4, Part 1, pp. 55–68.

TAYLOR, G. R. 1976. Styles of mineralisation in the Solomon Islands—a review. Pp. 83–91 in Marine geological investigations in the southwest Pacific and adjacent areas. G. P. GLASBY and H. R. KATZ (Editors). *Tech. Bull. Comm. Coord. Jt. Prospect. Miner. Resour. South Pac. Offshore Areas (CCOP/SOPAC)*, No. 2.

TERCIER, J. 1940. Dépôts marins actuels et séries géologiques. [Present-day marine sediments and geological rock series]. *Eclogae Geol. Helv.*, Vol. 32, pp. 47–100.

THAYER, T. P. 1967. Chemical and structural relations of ultramafic and felspathic rocks in Alpine intrusive complexes. Pp. 222–239 in *Ultramafic and related rocks.* P. J. WYLLIE (Editor). (New York: Wiley.)

THOMPSON, J. E. 1967. A geological history of eastern New Guinea. *J. Aust. Pet. Explor. Assoc.*, Vol. 7, Part 2, pp. 83–93.

THOMPSON, R. B. M. 1958. The geology of the Florida Group, 1956. Pp. 97–101 *in* The Solomon Islands—geological exploration and research, 1953–1956. *Mem. Geol. Surv. Br. Solomon Isl.*, No. 2, 151 pp.

— 1960. The geology of the ultrabasic rocks of the British Solomon Islands. Unpublished thesis presented for the degree of Doctor of Philosophy in the University of Sydney.

— 1965. Copper: geological and magnetic surveys of Hidden Valley and the southern flanks of Mt Gallego, western Guadalcanal, 1962. *Br. Solomon Isl. Geol. Rec.*, Vol. 2 (1959–1962), pp. 132–134.

— 1968. South-west Guadalcanal—the Itina River Basin, 1964–1965. *Br. Solomon Isl. Geol. Rec.*, Vol. 3 (1963–1967), pp. 9–14.

— and PUDSEY-DAWSON, P. A. 1958. The geology of eastern San Cristobal, 1955–1956. Pp. 90–95 *in* The Solomon Islands—geological exploration and research, 1953–1956. *Mem. Geol. Surv. Br. Solomon Isl.*, No. 2. 151 pp.

THORNBURY, W. D. 1954. *Principles of geomorphology.* 618 pp. (New York: Wiley.)

TILLEY, C. E. 1926. On garnet in pelitic contact zones. *Mineral. Mag.*, Vol. 21, p. 34.

TITLEY, S. R. 1975. Geological characteristics and the environment of some porphyry coppers in the south-western Pacific. *Econ. Geol.*, Vol. 70, pp. 499–514.

TURNER, C. C., EADE, J. V., DANITOFEA, S. and OLDNALL, R. 1977. (*In preparation*). Offshore gold prospecting, Guadalcanal, Solomon Islands.

— and HACKMAN, B. D. *In preparation.* The geology of the Beaufort Bay area, Guadalcanal. *Bull. Geol. Surv. Solomon Isl.*

TURNER, F. J. 1968. *Metamorphic petrology: mineralogical and field aspects.* 403 pp. (New York: McGraw-Hill.)

— and VERHOOGEN, J. 1951. *Igneous and metamorphic petrology.* 602 pp. (New York: McGraw-Hill.)

— and WEISS, L. E. 1963. *Structural analysis of metamorphic tectonites.* 545 pp. (New York: McGraw-Hill.)

VALLANCE, T. G. 1960. Concerning spilites. *Proc. Linn. Soc N.S.W.*, Vol. 85, pp. 8–52.

VAN DEVENTER, J. 1971. Geological reconnaissance survey, British Solomon Islands Protectorate. 15 pp. Unpublished report of Bataafse International Petroleum Maatschappij N.V., No. EP-41570 II.

WAGNER, L. R. and DEER, W. A. 1939. The petrology of the Skaergaard intrusion, Kangerdlugssuaq, east Greenland. *Medd. om Grønland*, Vol, 105, No. 4. 352 pp.

WALSHAW, R. B. 1974. A geochemical investigation of gold-bearing volcanic rudites at Gold Ridge, Guadalcanal. *Rep. Solomon Isl. Geol. Surv.*, 48 pp.

WARDEN, A. J. 1967. The geology of the central islands. *Rep. New Hebrides Geol. Surv.*, Vol. 5, pp. 1–108.

WARIN, O. N. 1969. Report on visit to Koloula River copper prospect, Guadalcanal, BSIP. Report of Utah Development Co., No. 160. 30 pp. [Unpublished.]

WELLS, T. 1965. Gold: The Kovagombi Area, Guadalcanal. *Br. Solomon Isl. Geol. Rec.*, Vol. 2 (1959–1962), pp. 103–107.

WENTWORTH, C. K. 1933. Fundamental limits to the sizes of clastic grains. *Science*, New Series, Vol. 77, pp. 633–634.

— and WILLIAMS, H. 1932. The classification and terminology of the pyroclastic rocks. *Bull. Nat. Research Counc.*, Vol. 89, pp. 19–53.

WESTWOOD, J. V. B. 1970. Seismicity of the Solomon and Santa Cruz islands, south-west Pacific. *J. Geol. Soc. Aust.*, Vol. 17, Part 1, pp. 87–92.

WHEELER, H. E. and MALLORY, S. V. 1956. Factors in lithostratigraphy. *Bull. Am. Assoc. Pet. Geol.*, Vol. 40, pp. 2711–2723.

WHITMORE, T. C. 1966. Guide to the forests of the British Solomon Islands. *BSIP For. Rec.*, Vol. 2. 208 pp.

— 1969. The vegetation of the Solomon Islands. *Philos. Trans. R. Soc. London*, Series B, Vol. 255, pp. 259–270.

WHITTEN, E. H. T. 1966. *Structural geology of folded rocks.* 678 pp. (Chicago: Rand McNally.)

WILLIAMS, A. 1959. A structural history of the Girvan District, S.W. Ayrshire. *Trans. R. Soc. Edinburgh*, Vol. 63, pp. 629–667.

WILLIAMS, H., TURNER, F. J. and GILBERT, C. M. 1954. *Petrography—An introduction to the study of rocks in thin sections.* 406 pp. (San Francisco: W H Freeman.)

WILSON, G. 1946. The relationship of slaty cleavage and kindred structures to tectonics. *Proc. Geol. Assoc.*, Vol. 57, pp. 263–302.

— 1961. The tectonic significance of small-scale structures and their importance to the geologist in the field. *Ann. Soc. Géol. Belg.*, Vol. 84, pp. 423–548.

WILSON, J. T. 1959. Geophysics and continental growth. *Am. Sci.*, Vol. 47, pp. 1–24.

— 1965. A new class of faults and their bearing on continental drift. *Nature, London*, Vol. 207, pp. 343–347.

WINKLER, H. A. 1968a. *Copper prospects of the Koloula River Granodiorite, Guadalcanal. UN Special Development Program Aerial Geophysical Survey Project Report.* (Honiara, Solomon Islands: Government Printing Office.)

— 1968b. *Regional geophysical structure of the British Solomon Islands. UN Special Development Program Aerial Geophysical Survey Project Report.* (Honiara, Solomon Islands: Government Printing Office.)

— 1968c. *Bauxite prospects of Rennell Island, British Solomon Islands Protectorate. Aerial Geophysical Survey Project, UNDP, British Solomon Islands Protectorate, 1965–1968.* 15 pp. (Honiara: Government Printer.)

WISE, D. U. 1963. Keystone faulting and gravity sliding driven by basement uplift of Owl Creek Mountains, Wyoming. *Bull. Am. Assoc. Pet. Geol.*, Vol. 47, pp. 586–598.

WOOLLARD, G. P., FURUMOTO, A. S., KROENKE, L., MALAHOFF, A., ROSE, J. C. and SUTTON, G. H. 1967. Cruise report on 1966 seismic refraction expedition to the Solomon Sea. *Hawaii Inst. Geophys. Rep.*, Vol. 67, Part 3.

WRIGHT, J. B. 1966. Convection and continental drift in the south-west Pacific. *Tectonophysics*, Vol. 3, Part 2, pp. 69–81.

WRIGHT, P. C. 1968a. Central Guadalcanal—the geology of the Ngalimbiu and Tangareso river systems. *Br. Solomon Isl. Geol. Rec.*, Vol. 3 (1963–1967), pp. 37–39.

— 1968b. Western Guadalcanal—the geology of the Lungga–Mbetikama and the Tenaru river systems. *Br. Solomon Isl. Geol. Rec.*, Vol. 3 (1963–1967), pp. 25–36.

YODER, H. S. and TILLEY, C. E. 1962. Origin of basalt magmas: an experimental study of natural and synthetic rock systems. *J. Petrol.*, Vol. 3, pp. 342–352.

List of co-ordinates of place names and topographic features

Name	Area of Guadalcanal	Latitude S	Longitude E
Aliemba River	SC	9° 47′	160° 09′
Alitauva River	SC	9° 46′	160° 12′
Alivaghato River	SC	9° 47′	160° 21′
Alualu River	SE	9° 53′	160° 33′
Alualumbo River	SC	9° 48′	160° 09′
Aola River	NE	9° 36′	160° 27′
Aola Bay	NE	9° 31′	160° 30′
Arandi River	EC	9° 48′	160° 37′
Arohaviha Lake	SC	9° 49′	160° 22′
Aruliho School and Village	NW	9° 17′	159° 46′
Asi River	SC	9° 47′	160° 25′
Avuavu Mission	SE	9° 50′	160° 23′
Bagarice Village	NC	9° 34′	160° 10′
Beaufort Bay	W	9° 37′	159° 39′
Cape Austen	SW	9° 42′	159° 42′
Cape Beaufort	W	9° 38′	159° 39′
Cape Henslow	SE	9° 56′	160° 38′
Cape Hunter	SW	9° 48′	159° 50′
Charichighi River	SC	9° 47′	160° 03′
Charihaivati Village	NC	9° 34′	160° 09′
Charihau River	NC	9° 31′	160° 12′
Charihaviha River	C	9° 43′	160° 08′
Charihoro River	C	9° 39′	160° 12′
Charikangge River	NC	9° 32′	160° 03′
Charikiki River	EC	9° 44′	160° 32′
Charikimberi River	C	9° 43′	160° 08′
Chariko Village site	C	9° 35′	160° 15′
Charilava River	SC	9° 46′	160° 02′
Chariturua River	NC	9° 34′	160° 33′
Chariveha River	NC	9° 34′	160° 12′
Charivunga River	NC	9° 35′	160° 08′
Chaunapaho (Jonapau) Mountain	NC	9° 38′	160° 07′
Choruchoru Gap	SC	9° 44′	160° 07′
Chovohio River	NC	9° 31′	160° 11′
Cone Peak Mountain	W	9° 27′	159° 42′
Esperance Mountain	NW	9° 15′	159° 43′
Gallego Mountain	NW	9° 21′	159° 43′
Ghalighecha River	SC	9° 47′	159° 58′
Gharangi Knoll	NC	9° 31′	160° 16′
Ghauregha River	NC	9° 31′	160° 15′
Ghecha River	SC	9° 46′	160° 03′
Ghoivara River	SC	9° 46′	160° 06′
Gold Ridge	NC	9° 36′	160° 08′
Haiacha Range	C	9° 42′	160° 03′
Hailava River	C	9° 44′	160° 20′
Haimarao Village	SE	9° 52′	160° 25′
Haimatua River	SE	9° 51′	160° 24′
Haimbau Village	C	9° 42′	160° 20′
Haimela Village site	C	9° 41′	160° 15′
Haiparia Syncline	C	9° 42′	160° 18′
Hairughu River	SE	9° 52′	160° 28′
Hanangga River	SE	9° 53′	160° 37′
Haviha Ridge	C	9° 38′	160° 13′
Hoilava River	W	9° 25′	159° 40′
Honiara	N	9° 25′	159° 58′
Hovu River	C	9° 41′	160° 22′
Humbaro River	EC	9° 38′	160° 24′
I Mbosa River	E	9° 45′	160° 43′
Isipeni Village	EC	9° 39′	160° 30′
Itina River	SW	9° 44′	159° 51′
Kaichui Mountain	SE	9° 46′	160° 30′
Kaimbaghasu River	SC	9° 47′	159° 58′
Kalangali	EC	9° 39′	160° 25′
Kandaho River	SE	9° 53′	160° 41′
Kaoka (Kaukau) Bay	E	9° 41′	160° 41′
Kavahambe (Kovagombi)	NC	9° 32′	160° 11′
Keli Ngali Hill	SE	9° 56′	160° 37′
Kolaninggeu River	SE	9° 54′	160° 33′
Kolilua River	SE	9° 54′	160° 38′
Kolitoa River	EC	9° 44′	160° 27′
Kolochoro River	SC	9° 47′	160° 02′
Kolochulu Village	C	9° 37′	160° 18′
Kolopura River	SE	9° 54′	160° 40′
Kologhailava River	SC	9° 47′	159° 59′
Kolohaisava River	EC	9° 42′	160° 33′
Kolokiki River	SE	9° 50′	160° 31′
Kolokokua River	EC	9° 38′	160° 28′
Kolokoo River	SC	9° 46′	160° 01′
Kolokuau River	SC	9° 46′	160° 03′
Kolombolavu River	C	9° 41′	160° 20′
Kolonggao River	EC	9° 41′	160° 36′
Kolopao River	EC	9° 46′	160° 37′
Kolotambu Swamp	SE	9° 50′	160° 23′
Koloula River	SC	9° 47′	160° 04′
Kolovaghamela River	E	9° 45′	160° 43′
Kolovagharindi River	NE	9° 37′	160° 22′
Kombito River	NE	9° 35′	160° 32′
Kombusoe area	NC	9° 34′	160° 12′
Komukama Village	NC	9° 31′	160° 18′
Kondokachi Point	SC	9° 49′	160° 09′
Kopiu Bay	SE	9° 53′	160° 46′
Korasahalu Reef	SE	9° 55′	160° 25′
Ko Ravu River	NE	9° 35′	160° 31′
Kova River	NC	9° 34′	160° 11′
Kuliamachapa Point	SC	9° 48′	159° 59′
Kuma River	SC	9° 46′	160° 10′
Lame River	NC	9° 36′	160° 17′
Lamulaghi River	SW	9° 43′	159° 49′
Latinarau Mountain	C	9° 42′	160° 10′
Lauvi Point and Lagoon	SE	9° 53′	160° 26′
Lee's Lake	EC	9° 42′	160° 24′
Lughumboko River	EC	9° 42′	160° 37′
Lungga Point	N	9° 24′	160° 02′
Makarakomburu Mountain	SC	9° 43′	160° 02′
Malagheti coastal area	SC	9° 49′	160° 07′
Malango area	WC & C	—	—
Mamasa River	NC	9° 32′	160° 12′
Manauvo River	SE	9° 53′	160° 42′
Mandonu River	E	9° 48′	160° 45′
Maotapuku River	NC	9° 37′	160° 07′
Marapa Island	E	9° 48′	160° 52′
Marau Sound	E	9° 48′	160° 51′
Maruiapa Island	E	9° 47′	160° 49′
Mataniko (Metanakau) River	N	9° 27′	159° 57′
Matepono (Metapona) River	N	9° 26′	160° 11′
Mauvo Raha gorge (Kolokumaha River)	C	9° 40′	160° 14′

Name	Area of Guadalcanal	Latitude S	Longitude E	Name	Area of Guadalcanal	Latitude S	Longitude E
Mavo River	NW	9° 22'	159° 50'	Pina River	SW	9° 40'	159° 46'
Mavugho gorge				Pinau Mountain	SE	9° 48'	160° 35'
(Mongga River)	EC	9° 39'	160° 23'	Poha River	NW	9° 25'	159° 53'
Mbahara River	EC	9° 43'	160° 32'	Poleo area	SW	—	—
Mbalanga River	NC	9° 30'	160° 20'	Popomanaseu Mountain	C	9° 42'	160° 04'
Mbalisuna River	N	9° 29'	160° 13'	Popori Mountain	NW	9° 20'	159° 42'
Mbalo Village	SE	9° 56'	160° 38'	Pukarauvaghalo Mountain			
Mbambakavoa River	SC	9° 44'	160° 05'	range	SC	9° 43'	160° 02'
Mbarona River	WC	9° 33'	159° 54'	Purasakusu River	EC	9° 38'	160° 29'
Mbeambea River	C	9° 38'	160° 03'	Reko area	N	9° 28'	160° 21'
Mbeghovila River	NC	9° 33'	160° 15'	Rere River	NE & EC	9° 37'	160° 33'
Mbelatania Point	SE	9° 55'	160° 33'	Rerembonumbonu River	NE	9° 36'	160° 35'
Mberande River	N	9° 30'	160° 17'	Rere Point	NE	9° 34'	160° 37'
Mberina River	NE	9° 34'	160° 30'	Riva River	SC	9° 47'	160° 12'
Mbetilonga Village	NC	9° 33'	159° 59'	Roundhead Mountain	NW	9° 16'	159° 43'
Mbetisahata River	WC	9° 33'	159° 56'	Rua Sura Island	—	9° 30'	160° 37'
Mbetivatu River	NC	9° 31'	160° 05'	Sambahalava River	SE	9° 50'	160° 32'
Mbicho River	C	9° 37'	160° 01'	Sambaharihi River	SE	9° 50'	160° 29'
Mbina River	SC	9° 44'	160° 03'	Sambahatangi Mountain	C	9° 42'	160° 04'
Mbokokimbo River	NE	9° 30'	160° 23'	Sani Lumu hamlet	NE	9° 35'	160° 34'
Mbolavu River	SE	9° 49'	160° 25'	Sarivuro River	NC	9° 32'	160° 19'
Mbolonda Bay	SW	9° 47'	159° 53'	Simiu River	NE	9° 38'	160° 39'
Mbonehe River	NW	9° 24'	159° 50'	Sisi River	SC	9° 48'	160° 02'
Mbongialo River	C	9° 42'	160° 08'	Soto River	NE	9° 34'	160° 24'
Mbo'o River	E	9° 42'	160° 39'	Sukiki Village	SE	9° 56'	160° 37'
Mboranaho Village	EC	9° 43'	160° 28'	Sura Kiki Island	—	9° 30'	160° 38'
Mborambora River	C	9° 43'	160° 18'	Susu River	NE	9° 33'	160° 33'
Mbuha River	NE	9° 35'	160° 29'	Sutakama River	C	—	—
Mbulo River	NE	9° 36'	160° 36'	Sutakiki River	C	9° 40'	160° 07'
Mbumbunuhu Village	EC	9° 38'	160° 23'	Sutalanga River	SW	9° 43'	160° 00'
Mburumburu River	E	9° 44'	160° 43'	Talise River	SC	9° 47'	160° 16'
Mount Austen	N	9° 29'	159° 59'	Talise Anchorage	SC	9° 48'	160° 16'
Na Asa Gorge	C	9° 39'	160° 20'	Tambunanguu Mountain	WC	9° 39'	159° 59'
Na Hoeta Lake	EC	9° 45'	160° 36'	Tanareirei Mountain	SC	9° 43'	160° 05'
Na Hue River	EC	9° 43'	160° 35'	Tangarare River	W	9° 33'	159° 41'
Na Humbu River	E	9° 46'	160° 41'	Tangareso River	NC	9° 32'	160° 03'
Na Keo River	SE	9° 51'	160° 44'	Tanggiata River	SE	9° 50'	160° 26'
Na Koilo River	EC	9° 38'	160° 27'	Tangisi Koivo Lake	SC	9° 46'	160° 12'
Na Kuma = Lee's Lake	EC	9° 42'	160° 24'	Tatuve Mountain	C	9° 40'	160° 10'
Namokanji River	SC	9° 46'	160° 21'	Tausoro River	NC	9° 35'	160° 10'
Namonambosa Village	C	9° 40'	160° 16'	Tavangaoa River	NE	9° 37'	160° 37'
Namondaula River	SC	9° 47'	160° 24'	Tawa'ihi Island	E	9° 50'	160° 50'
Na Ndoli River	EC	9° 41'	160° 35'	Tenaru River	N	9° 27'	160° 04'
Na Papa River	EC	9° 45'	160° 34'	Tetekanji area	EC	9° 47'	160° 36'
Na Puleo River	E	9° 49'	160° 42'	Tetevasa River	EC	9° 39'	160° 27'
Na Suha Mountain	SE	9° 49'	160° 36'	Tevua Village	SC	9° 42'	160° 20'
Na Vuvuti Village	EC	9° 46'	160° 36'	Tina River	NC	9° 32'	160° 05'
Na Uli River	EC	9° 42'	160° 31'	Tinahulu River	NC	9° 30'	160° 09'
Ndatovitu Mountain	NC	9° 36'	160° 12'	Tinggetingge Mountain	C	9° 44'	160° 11'
Nduindui Village	SC	9° 48'	159° 58'	Tinomeat Village	NC	9° 35'	160° 08'
Ngalimbiu River	N	9° 29'	160° 08'	Tiraone River	EC	9° 44'	160° 29'
Nggeunaha River	C	9° 40'	160° 15'	Toghasa Mountain	EC	9° 45'	160° 25'
Niumakani Village site	EC	9° 39'	160° 29'	Toghonani	SC	9° 48'	160° 07'
Njarikusu River	NE	9° 35'	160° 23'	Toni River	NC	9° 33'	160° 07'
Njarivila River	NC	9° 35'	160° 14'	Tualoto River	C	9° 43'	160° 06'
Nudha Island	NE	9° 31'	160° 48'	Tumanggulu Village	NE	9° 36'	160° 36'
Nuhu Village	C	9° 38'	160° 10'	Tutumu Village	N	9° 29'	160° 16'
Oa River	SE	9° 53'	160° 44'	Uluakolo River	SC	9° 44'	160° 21'
Oa Village	SE	9° 54'	160° 44'	Umasani River	NW	9° 21'	159° 49'
Old Case Village	NC	9° 33'	160° 10'	Unggeva River	SE	9° 52'	160° 37'
Paluluna doline	E	9° 49'	160° 44'	Vaghanambo River	EC	9° 44'	160° 26'
Paripao area	NE	—	—	Vaghanambo Gorge	EC	9° 40'	160° 28'
Paru Mountain	NW	9° 19'	159° 41'	Vaitambu River	E	9° 44'	160° 41'
Paruru Village	E	9° 51'	160° 50'	Valasi River	EC	9° 40'	160° 32'

Name	Area of Guadalcanal	Latitude S	Longitude E
Vale Muse Village site	EC	9° 41′	160° 27′
Valenakoko Village site	EC	9° 41′	160° 29′
Valevuru Village	SC	9° 47′	160° 03′
Vatukulau sea stack	SC	9° 49′	160° 08′
Vatulava River	SE	9° 53′	160° 43′
Vatupochau Mountain	EC	9° 41′	160° 26′
Vaumandali River	E	9° 45′	160° 42′
Veuru Moli area	SE	—	—
Visale Mission	NW	9° 15′	159° 42′
Viso River	SC	9° 47′	160° 00′
Vui River	C	9° 38′	160° 17′
Vuraka River	EC	9° 41′	160° 24′
Vurakirapa River	C	9° 42′	160° 06′
Wahere Island = Komachu	E	9° 49′	160° 49′

Note: Letter 'C' in the second column denotes 'central'.

INDEX

ABEM, 4, 47, 85
Aerial photographs, 7
Age determinations, 17–18, 43–44
Agriculture, 9
'Albatros' expedition, 3
ALBERS, J. P., 99
Algal concretions, 29
Aliemba River, 29
Alitauva River, 15
Alite
      Limestones (Malaita), 92
      Volcanics (Malaita), 91
Alivaghato River, 52, 59, 66–67, 71
ALLUM, J. A. E., 4, 8, 33, 36, 42–43, 60
Almandine, 27
Alualu River, 18, 64
Alunite, 75
AMHERST, Lord, of Hackney, 3
Amphibolite facies, 27
AMSTUTZ, G. C., 24
ANDERSON, E. M., 68
Andesites, 44, 90
Anorthosite, 24, 26
Anuha Calcarenite (Florida), 91
Aola River, 3, 18, 37, 41, 60
Apatite, 75
Aquitanian, 33
Ara, 10
Arohaviha Lake, 15
Arsenic, 73–76
Arsenopyrite, 76
Aruliho, 43
Atoll Province, 3
AUBOUIN, J., 100
Avuavu, 8, 15

BADGLEY, P. C., 64, 101
Bagarice, 41
Balasuna Syndicate, 76
'Ball-and-pillow', 35
BARRUS, R. B., 73–74
Balsaltic andesites, 29, 41, 88, 97
Basalts, 19, 30
Basement Complex, 4, **17**
Bauxite, 77–78
Beach Sands, 3, **78**
Beaufort Bay, 4, 25, 78
Bedding,
      in the Mbirao Volcanics, 48–49
      in the Neogene Sequence, 56–59
BELL, J. D., 91–92, 94
BELOUSSOV, V. V., 94
BENSON, W. N., 100
Berande Beds, 4–5, **43**
BERSENEV, I. I., 30

Betilonga Limestone, see Mbetilonga, 4–5, 33
BILLINGS, M. P., 66
Bioturbation, 35
Bismarck
      Plate, 101
      Volcanic Arc, 1, 101
Bivalves, 36
BLAKE, D. H., 91–92
Blanche Channel, 89
Bokokimbo Beds, see Mbokokimbo, 5
Bonegi Limestone, see Mbonehe, 4, 33
Bornite, 73–75
Bougainville, 1, 91–92
Bruchfalten, 89
BUDDINGTON, A. F., 25
Buka, 1
Bulolo Gold Dredging Co. Ltd, 77
Burdigalian, 33
Bytownite, 27

CADY, W. M., 100
Calc-alkali rock series, 44–45, 97
Calcareous tuff, 47
Cape Austen, 12
Cape Beaufort, 10, 12, 28
Cape Esperance, 4
Cape Henslow, 3, 11, 15, 18, 43, 46, 78
Cape Hunter, 12
Carbonaceous schist, 22
CAREY, S. W., 67, 89, 93, 99–101
CAROZZI, A. V., 35
Cataclastic cleavage, 52
Caves, 17
Cementstone, 78
Central Province, 1, 87, 90, 100–101
Chalcocite, 73–75
Chalcopyrite, 41, 71, 73–75
Charichighi River, 45
Charihaivati, 60
Charihau River, 39
Charihaviha River, 27
Charihoro River, 39
Charikangge (Charikange) Beds (and Grit), 3, **38–40,** 89
Charikiki River, 59
Charikimberi River, 16, 19
Charikumbau River, 73
Charilava River, 44
Chariturua River, 39
Chariveha River, 41
Charivunga River, 39 ,41
Charters Towers lineament, 100
Chaunapaho Mountain, 9
CHAYES, F., 24
Chembea River, 73
Chemical analyses, 23–25, 30–32, 44–45, 78, 97
Chert, 19
Chikora River (Koloula), 74
Chimiu
      Fault Zone, 61, 86
      syncline, 60
CHIVAS, A. R., 74
Choiseul Schists, 87–88, 91
Choruchoru Gap, 10

Chovohio
      Igneous Phase, 41
      Island, 89
      (Sorvohio) River, 3, 35, 39–41, 60, 76, 80
Chromite, 78
Climate, 10
COATS, R. R., 97
Cobalt, 73
'Cockpit karst', 17
COLEMAN, P. J., 1–5, 7, 17–18, 25, 32–33, 35–36, 38–39, 41–43, 46–48, 58–60, 63–64, 67, 77, 87, 89–93, 100–102
COLES, I. G., 73
Communications, 8
Cone Peak, 11
CONNOLLY, H. J. C., 76
CONNOLLY, J. W., 3, 78
CONYBEARE, C. E. B., 33, 38
Copper, 45
      in oceanic tholeiites, 71–73
      in felsic igneous association, 74–76
'Copper Province', 72
Coral reef, 48
COTTON, C. A., 15
Covellite, 75
CRAIG, P., 2, 73, 92
Crenulation foliation, 52
Cretaceous, 87
CROOK, K. A. W., 33, 38
CULLEN, D. J., 93, 101
CURTIS, J. W., 85, 101
'Cushion' topography, 25
'Cut and fill' structures, 38

DALY, R. A.. 97
Darwin Rise, 101
DAVIES, H. L., 91, 97
DAY, A. A., 38, 43, 77, 79, 80
DE GOLYER and MACNAUGHTON, Inc, 5, 77
DENHAM, D., 84, 101
DENNIS, R. A., 4
DE SITTER, L. U., 67, 93
DEWEY, J. F., 88
Dextral faulting, 63–64, 86, 88
Diaspore, 75
DIETZ, R. S., 91
Discontinuity, geophysical, 86
Diseases, 9
DU FAUR, B., 77
Dunite-serpentinite, 27
Dykes, 20–21, 27, 41, 44, 67–69

EDWARDS, A. B., 42, 91
Energy index, 35
ENGEL, A. E. J. and ENGEL, C. G., 87
Enstatite-gabbro, 27
Eocene, 87–88
Epidosites, 22
Esperance Mountain, 11

FAIRBRIDGE, R. W., 93
Fauna, 9–10
Faulting, 60–66
Ferro-femic index, 97
First Geological Map of the Solomons, 3

FISHER, R. V., 30
Fiu Lavas (Malaita), 91
Flaser rocks, 22
Florida Islands, 78
Fluid inclusions, 74
Folding, 58, **59–60**, 94
FOLK, R. L., 34
Foraminifera, 29, 33, 35–36, 39, 46

Gallego
    Lavas, 4, **43**, 92
    Mountain, 11, 43, 75
    Volcanic Zone, 11, 42, **43**, 85
GARDNER, M. E., 72, 75
Garnet-mica-schist, 27
Garnierite, 74
GASS, I. G., 91
Gastropods, 33
Geographical names, 9
Geomorphology, 10
Geothermal springs, 46
Ghalighecha River, 30
Gharangi Knoll, 47, 85
Ghari, 11, 73
Ghauregha River, 46–47
Ghausava
    Fault Zone, 61, **66**, 89
    River, 73, 77
Ghausava (Gausava) Ultrabasics, 3, 5,
    21, **27**, 58, 67, 73, 80, 86, 89, 94
Ghecha River, 44, 67, 74
Ghoivara River, 44
Ghunukeo River, 73
GLAESSNER, M. E., 90, 93
Globorotalia spp., 18
Glomeroporphyritic feldspar dolerites,
    21, 68, 87
Goethite, 75–76
Gold, 3, 41, 48, 73–74, **76–77**, 89
Gold Ridge, 3–4, 66, **76**, 80, 85, 89
    Volcanics, **38–41**, 67, 69, 73
GRASSO, V. G., 24, 30
Gravity, 79–83
GREEN, D. H. and PITT, R. P. B., 100
Greenschist facies, 23
GROVER, J C , 3–5, 16, 19, 40, 70–71,
    73–79, 85, 100
'Grush', 44, 67
Guadalcanal
    Basement, 5
    Gabbro, **21**, 68, 87, 88
    Igneous Complex, 4
    Plains, 9, 11
    Schists 5, 22
GUPPY, H. B., 3
Gypsum, 73

HACKMAN, B. D., 5, 28, 46, 73–76, 78,
    92, 94, 99
Haiacha, 7, 10, 16, 28, 81
Haimarao, 8
Haimatua River, 15
Haimbau, 36
Haimela, 19, 59, 60
Haiparia syncline, 60
Hairughu River, 20
Halimeda, 33, 35, 42

HALUNEN, A. J. and HERZEN, R. P. VON,
    102
HAMMOND, B., and GOSS, B., 78
Hanangga
    Fault, 61
    River, 15, 18
Harigha Conglomerates (San Cristobal),
    91
Harzburgite, 26–27
HATCH, F. H., 44
Haviha
    Ridge, 11, 47, 90
    Sandstone, 39
Hawaiian Oceanic Olivine-Basalt
    Association, 25
Hematite, 19, 72, 75–76, 78
HESS, H. H., 99
Hidden Valley, 75–76
High Alumina Basalt Series, 30
High-point strip profiles, 11
HILL, J. H., 3, 32, 58, 75
HILLS, E. S., 64
HODGSON, J. H., 85
Hoilava River, 58, 76
Holocene, 90
Honiara, 4, 8
    Beds, 4, 11, 17, **46–47**, 78, 90, 92
    Reef Limestones, 47, 77–78
Horohana, 11
Hovu River, 33
HUGHES, G. W., 4, 7, 42, 75
Humbaro River, 36
HUZITA, K., 85, 100
Hyaloclastite, 19
Hybridisation, 45
Hypersthene-andesite, 40–41, 89
Hypersthenic Series (Japanese), 30, 44,
    92

ILLIES, J. H., 100
Ilmenite, 78
I Mbosa River, 26–27
Indispensable Strait Basin, 79–80
Ingrown meanders, 16
Iron, 78
ISACKS, B. L., 85
Isipeni Limestones, 5, 34
Isostatic compensation, 81
Itina
    Fault, 66
    River Basin, 3–4, 11, 29, 32–33, 58,
    60, 67, 77, 88

JAKES, P. and WHITE, A. J. R., 90, 92,
    97, 99
Jarosite, 76
JARVIS, D. M., 72
JOHANNSEN, A., 26–27, 45
JOHNSON, T. and MOLNAR, P., 101
Jointing, 66–67, 86
JONES, O. A., 76
JOPLIN, G. A., 25, 45, 90

Kaichui Mountain, 11
Kaimbaghasu River, 30
Kaipito–Korighole Thrust (Santa
    Isabel), 99

KAJEWSKI, 3, 76
Kamangga Grit (Choiseul), 91
Kaoka (Kau Kau Bay), 9, 11, 46, 86
KARIG, D. D. and MAMMERICKX, J., 102
Karst labyrinth, 17
Kaukau Lavas, 5
Kavahambe (Kovagombi), 3, 48, 77
Kavo
    Greywacke Beds, 3, 5, 29, **32**, 67, 88,
    91
    Ranges, 3, 7, 10, 32, 61, 81, 90
    Shales, 3, 32
    Siltstones, 3, 32
KAY, M., 100
Keimane River, 77
Keli Ngali, 11
Keriaka Limestone (Bougainville), 91
Keystone graben, 93
KNILL, J. L., 52
KNOPF, E. B., 52
Kolaninggeu River, 6
Kolochulu, 11, 47
Kologhailava River, 30
Kolohaisava
    Mudstone, 5, **36–37**, 90
    River, 9, 15, 49, 72
Kolokiki River, 18
Kolokokua River, 17
Kolokoo River, 16, 44
Kolokumaha River, 22, 34, 37, 39
Kolombolavu
    Fault, 60
    River, 33, 36–38, 52, 59, 71–72
Kolonggao River, 35
Kolopao River, 22
Kolopura River, 20
Kolotahombui River, 77
Kolotambu swamp, 15
Koloula
    Diorite (Granodiorite) Complex, 4, 29,
    44–45, 64, 67, 69, 90, 92, 99
    River, 3, 11–12, 16, 28–29, 44–45,
    77
Kolovaghamela River, 19, 22, 26–27,
    49, 59, 86
Kolovagharindi River, 36
Kombito River, 9, 16, 42, 73
Komburu, 10
Kombusoe Limestones 39
Komukama, 9
Kopiu Bay, 46, 49
Korasahalu Reef, 15
Ko Ravu, 42
Kosikosi River, 74
Kova River, 40–41
Kovagombi, *see* Kavahambe
KRAUSE, D. C., 93, 100
KRUMBEIN, W. C. and SLOSS, L. L.,
    42, 89
Kuma Valley, 10, 22, 27, 29–30, 61–68
Kumboro Volcanics (Choiseul), 92
KUNO, H., 44, 92, 94, 99

Lacustrine deposits, 48
LADD, H., 33, 39
Lake Lee Calcarenite, 4–5, **33–35**, 36,
    60, 88–90

Lame River, 36, 66
Lamulaghi River, 32
Lapeti River, 77
Latinarau Mountain, 16
LAUBSCHER, H., 100
LAUDON, T. S., 4, 79–81
Lauvi
    Lagoon, 15
    Point, 12
Lee's Lake, 11, 17, 33, 61
Lignite, 77
Lihir Islands, 1
Limonite, 75, 78
LINDEN, W. J. M. VAN DER, 100–101
Lineaments, 48, 60
Longgu, 77
    Beds, 5, **46**
Lotu (Mbetilonga), 75
Louisiade Sphenochasm, 101
Lughumboko River, 9, 18, 22
Lunga Volcanics, 4, 42
Lungga
    Beds, 3, 32, **42–43,** 58, 78
    Plateau, 11, 15
    River, 3, 9, 16, 42, 58
    shelf, 80
    valley, 4
LUYENDYK, B. P., 101

McBIRNEY, A. R., 97
MACDONALD, G. A., 23, 94
MACDONALD, G. A. and KATSURA, T., 94
McKEE, E. D., 35, 38
MACKENZIE, D. H., 72, 75
McKENZIE D. P. and MORGAN, W. J., 100
McKINSTRY, H. E., 64
McTAVISH, R. A., 3, 5, 7, 35, 37, 39, 41, 89
Maetambe Volcanics (Choiseul), 92
Magnetic survey, 85–86, 100
Magnetite, 71, 72, 78, 88
Makarakomburu Mountain, 10, 28
Malachite, 41, 73–75
Malagheti, 12, 29
MALAKHOV, I. A., 28
Malango, 22, 90
    Fault, 29, 61, 80
Malekula, 48, 92
MALLICK, D. I. J., 91
Mamasa River, 39–40
    Volcanics (New Hebrides), 91
Manauvo River, 15
Mandonu
    Fault, 19, 22, 61
    River, 17, 22, 49, 52–53
Manganese, 20, 73, 75–76
Manuhoho River, 15–16, 21, 36, 60
MARANZANA, F., 2, 4, 35, 91
Marapa Island, 11, 22
Marasa
    Tuffs, 5
    Volcanics, 28, 88
Marau, 7, 11–12, 15, 19, 22, 48, 52
    Ultrabasics, 3, 5, 26, 52, 63, 74, 80, 85, 88, 93
MARSHALL, C. E., 27

Maruiapa Island, 25
MASASHIGE, H., 67
MASON, D. R., 75
Mataniko
    River, 3
    Tuffs, 43
Matepono River, 3, 39, 76–77, 80
Mauvo Raha gorge, 34
Mavo River, 8, 66, 74–75, 77
Mavugho gorge, 37–38
Mbahara River, 25
Mbahomea Ridge, 89
Mbalanga
    River, 37
    Shale, **36–38,** 66, 90
Mbalisuna
    Discontinuity, 41
    Fault Zone, 61, 80, 86
    Gabbro, **41,** 45, 47, 67, 80, 85, 90, 99
    River, 16, 27, 80, 85
Mbalo River, 46
Mbeambea
    Fault (Thrust), 89, 93
    Valley, 22, 88
Mbeghovila, 41
Mbelatania Point, 15
Mberande, 7
    Beds, 37, **43**
    River, 16, 28, 77, 85
    Tongue, 37, 39, 89
Mberina River, 42
Mbetilonga, 3, 4, 17, 33, 58, 80, 88
    Basin, 11, 22, 75, 85
    Group, **33,** 91
    Limestone, 29, **33–35,** 58–59, 88–89
Mbetisahata River, 32, 75
Mbetivatu
    River, 39
    Sandstone, **38–40,** 90
Mbicho River, 21
Mbina River (Koloula), 74
Mbirao
    Block, 18, 80, 83, 100
    Dolerites, **20,** 87
    Geanticline, 87
    Group, **17,** 64, 71, 85, 90, 97
    Lavas, 5
    Mafic Igneous Suite, 73
    Metabasics, **22–23,** 52, 72, 88
    Mountains, 10
    Valley Zone, 11, 15, 22
    Volcanics, **18,** 49, 87
Mbokokimbo
    Basin, 38, 77, 85, 89, 100
    Beds, **36,** 60
    Formation, **36–37,** 43, 67
    Lithosome, 37–38, 88
    River, 11, 15–16, 28, 77, 80
    shelf, 80
Mbolavu River, 15, 20, 60
Mbolonda Bay, 28
Mbombo River, 73
Mbonehe
    Limestone, 17, 31, **33–34,** 58, 77–78, 88
    River, 9
Mbongialo River, 27

Mbo'o River, 46
Mborambora Thrust, 66, 93
Mboranaho, 19
Mbota Moli, 15
    Beds, 5, 18, 43, **46–47**
Mbuha River, 42
Mbulo River, 9
Mbumbunuhu, 38
Mburumburu River, 24
Mbusasangatu River, 75
Melanesian
    Border Plateau, 93
    Re-entrant, 93, 100
    Rise, 93
    Shear Zone, 100–101
Mellish Rise, 100
MENARD, H. W., 93
MENDAÑA, A. DE, 3, 76
Mendrausuthu Andesitic Group (Fiji), 92
Metanakau Tuffs, *see* Mataniko, 4
Metapona Beds, 4
MIEZITIS, Y., 91–92
MILSOM, J. S., 101
Miocene, 88, 91
Misfit streams, 15–16
MITCHELL, A. H. G., 49, 92
MIYASHIRO, A., 27
Mole Formation (Choiseul), 91
Monazite, 78
Mongga River, 17, 33, 36–38, 67, 77, 86
MOODY, J. D., 93
MOODY, J. D. and HILL, M. J., 64, 69
Mt Austen Beds, 4, 5, 42, **43**
Mt Vasu Limestone (Choiseul), 91
MURATA, K. J., 24

Na Asa Gorge, 38
Na Hoeta Lake, 17
Na Hue River, 17–18, 20
Na Humbu River, 59
Na Ndoli River, 34, 60
Na Papa River, 59
Na Puleo River, 19
Na Uli River (Valasi), 48
Na Vuvuti, 20, 73
Ndatovitu Mountain, 11, 47
Ndoma, 4
Nduindui, 7, 12, 30, 61
Nepheline-trachyte, 30
NEWMAN, D., 73
NEWMAN, D. and YOULES, I., 75
Ngalimbiu
    Alluvials, 4–5, **48**
    River, 3, 8
Nggae, 78
Nggela Group, 91
Nggeunaha River, 15–16, 18–19, 52, 60
Nickeliferous laterites, 73
Njarikusu River, 38
Njarivila River, 41
Nudha Island, 15, 48
Nughu Island, 80
Nuhu, 16, 27
Nukiki Limestone (Choiseul), 92

Oa, 3, 15, 18, 78
    Fault, 61
    River, 19–20

Oblique-slip faulting, 56
Oceanic Volcanic Province, 2
OFFICER, C. B., 93
Oil Search Ltd., 77
Old Case, 41
Oligocene, 87–88
Olistoclasts, 38
Olivine-andesite, 30
OLSEN, A., 75
Omboombo, 76
Ontong Java Rise, 1
Operculina complanata, 35
Ophiolitic complexes, 91
Orocline, 93, 99
O'SULLIVAN, K. N., 70, 72–73

Pacific Province, 2, 87, 100–101
Palaeocene, 87
Palagonite, 41
Paluluna, 17
Panguna, Bougainville, 92
Paraconglomerates, 40
Paripao
    Discontinuity, 86
    Tongue, 37, 39, 41, 90
Paru–Popori mountains, 11
Paruru, 8, 15
PEACOCK, M. A., 24, 30
Pebble dyke, 74
Pelowou Volcanics (New Hebrides), 91
Pemba Siltstones (Choiseul), 92
Pentecost Island, 91
Peteao Limestone (New Hebrides), 91
Petrochemical evolution, 94
Petroleum, 77
PETTIJOHN, F. J., 33, 35, 39
Photogeology, see ALLUM, J. A. E., 48
Pina River, 28
Pinau–Na Suha Range, 11
PISARSKI, J. P., 72, 77–78
Planet Deep, 1, 84
Planktonic foraminifera, 38, 42, 77, 89, 91
Plate tectonics, 101–102
Pleistocene, 90
Pliocene, 89–90
PLUMLEY, W. J., 35
Plunge of fold axes, 59–60
Pocklington Fault, 93, 100–101
Poha
    Diorite, 30–31, 74, 91
    River, 4, 28, 74, 88
POLDERVAART, A., 23
Poleo, 11
Popomanaseu Mountain, 10. 28
Population, 9
Porphyrite, 41
Porphyry copper association, 74
PRINZ, M., 71
Protoclastic deformation, 21, 27
Ptygmatic folds, 26
PUDSEY-DAWSON, P. A., 3–4, 16–17, 19,
    28, 31, 34, 42–44, 58, 66–67, 74, 76,
    78, 91
Pukarauvaghalo, 10
Pumpellyite, 19, 23
Purasakusu River, 36
PURVIS, J. G., 76

Pyrite, 19, 41, 71–72, 74, 75–76, 88
Pyrrhotite, 19

Quartz, 41, 71–75

Radiometric anomaly, 47, 74, 77
Radium, 77
Ravo Limestones, 18, 91
REBEK, R. J. and THOMAS, G., 78
Recent Sediments, 48
Reko, 37
Rennell Island, 77
Rere River, 16, 18–21, 59, 73
Rerembonumbonu River, 16
Rere Point, 86
REVELLE, R. R., 20
Rhombochasms, 100
RICHARDS, J. R., 18, 88
RICKWOOD, F. K., 91–92, 94, 99
RIPPER, I. D., 85, 101
Riva River, 15, 61, 66
ROBINSON, G. P., 11, 91–92, 101
RODDA, P. and BAND, R. B., 91–92
ROSE, J. C., 4, 79–80, 85, 100
Roundhead Mountain, 11
Rua Sura Island, 3, 15, 48
Rutile, 78

Sambahalava River, 15, 18, 49, 61, 64
Sambaharihi River, 15, 20–21, 63
San Cristobal Trench, 1, 85, 90
Sanidine, 29
Sani Lumu, 16
Santa Cruz, 1, 2, 84, 92
    Basin, 1
Sarivuro River, 38
Savo Island, 43, 66
Savulei, 46, 80, 90
    Fault Zone, 43, 66, 67
Saxonite, 26
Schistosity, 52
Seismicity, 83–85, 90, 100–101
SHARFMAN, V. S., 24
SHARPE, C. F. S., 16
SHIDÔ, F., 27
Siderite spherulites, 37
Sigana Volcanics (Santa Isabel), 87, 91
Sills, 20
Simiu River, 9, 15, 41, 46
Sinistral faulting, 27, 61, 63–64, 82, 86,
    93
Siu River, 74
Slot Basin, 79, 85
SMITH, B., 35
SNELLING, N. J., 17, 43–44, 87
Soghonara Lavas, 91
Sohano Limestone (Bougainville), 92
Solomon Sea Plate (Solomons Plate), 101
Sorvohio, see Chovohio,
Sotokiki River, 39
Sphalerite, 74–75
Sphenochasms, 100
Spilite–keratophyre association, 24–25, 87
S-surfaces, 52–56, 63
STANTON, R. L., 3, 25, 48, 77, 91–92,
    94, 97, 99
Step-faulting, 61, 66

STODDART, D. R., 15, 48
Strike-slip faulting, 61–64, 85, 93
Structural geology, 92ff
Sukiki, 15
Suiphides, massive, 73
Sura Kiki Island, 15
Sura River, 74
Susu, 77
    River, 16
Suta Fault, 22, 29, 35, 61, 66, 69, 89
Suta–Horohana Ranges, 10–11
Sutakama River, 3, 15, 19–22, 25, 27,
    30, 35, 41, 49
Sutakiki
    Series, 4
    Valley, 27, 73
Sutalanga River, 29
Suta
    Ultrabasics, 22, 27, 52, 61, 74, 80, 85,
        88, 93
    Volcanics, 22, 28–31, 44, 56, 64, 69,
        88, 91
Syntaphral tectonics, 89

Tabar Islands, 1
Talc, 75
Talise, 4, 15, 71
    River, 11, 49, 52, 68, 72
Tambunanguu Mountain, 10
Tanaemba River, 78
Tanareirei Mountain, 11
Tangaraisu
    Beds, see Tangareso, 9
    Marl, 4, 35
    Shale, 5, 35
Tangarare River, 58, 76
Tangaresco
    Beds, 9, 35–36, 89
    Channel, 89
    River, 3, 39
Tanggiata River, 64
Tangisi Koivo, 15
Tanjili Mountain, 11
Taphrogeosynclines, 100
TARLING, D. H., 100
Tasimboko, 78
Tatuve Mountain, 3, 11, 48
Tausoro
    Anticlinal horst, 60, 76, 90
    Valley, 35, 39, 60, 66
Tavangaoa River, 9
Tawa'ihi Island, 11
Tawoli Calcarenites (New Hebrides), 92
TAYLOR, G. R., 77–78
Tectonic domains, 48
Tenaru
    Conglomerates, 4, 43
    River, 3, 17
Tenorite, 75
TERCIER, J., 89
Terraces, 47–48, 90
Tetekanji
    Fault, 61, 72–73
    Limestones, 17, 19–20, 49, 73, 87–88
Tetere Anomaly, 77, 80
Tethyan Torsion Zone, 93
Tevua Conglomerate, 35

THAYER, T. P., 99
Tholeiitic vulcanicity, 88
Tholo Volcanics (Fiji), 91
THOMPSON, J. E., 91–92
THOMPSON, R. B. M., 3, 5, 11, 17, 18,
   21–22, 25–29, 31–34, 42–44, 52, 58,
   63, 66–67, 73–75, 77–78, 87, 89,
   91–94, 99
THORNBURY, W. D., 16
Tiaro
   Bay, 4, 8, 75, 76, 78
   Tuff-Breccias, 75
TILLEY, C. E., 23, 27, 30
Tina
   Calcarenite, 4, 5, **33–34,** 58, 66
   flexure, 60
   River, 21, 28–29, 34, 60, 71, 73, 80
Tinahulu River, 29, 33, 35, 39, 58,
   60–61, 64
Tinggetingge Mountain, 11
Tinomeat, 60
Tiraone River, 22
Titania, 24–25
TITLEY, S. R., 78
Toghasa Mountain, 11
Toghonani Fault, 44
Tomba Silts (Malaita), 92
Tonalite, 44–45
Toni
   Beds, 38, 89–90
   Formation, **38–41,** 48, 60
   Lithosome, **39–40,** 76, 89
   River, 41
   Tuffs, 4, 38
Torbernite, 74
Torres Trench, 1, 85
Trachyandesite, 29
Transform fault, 85
Tridacna, 46
Troctolite, 27
Tsunamis, 89
Tualoto Limestones, 30
Turbidites, 33
TURNER, C. C., 4, 73–74, 77
TURNER, F. J., 23–25, 27, 48, 52, 59,
   88

Tutumu, 47

Ulawa Trench, 1
Ultrabasic rocks, 25–28
Umasani
   River, 28, 43, 74–75
   Tuff(s), 4, **43**
   Volcanics, 28, 75
Uraghai River, 77
Uranium, 74

Vaghanambo River, 9, 17–18, 20–22,
   33–37, 68, 73
Vaitambu River, 26
Valasi
   Block, 18, 22, 34, 49, 52, 61, 68, 73,
     85, 87–88
   Fault Zone, 18, 61, 73, 86
   Limestones, 5, **33–35,** 90
   Mafic Igneous Suite, 73
   River, 9, 18, 37, 48
Vale Muse, 68, 73
VALLANCE, T. G., 24
VAN DEVENTER, J., 77
Vara Creek, 46
Vatukulau, 12
Vatulava River, 15
Vatumbulu Beds, 5, 37, **41–42,** 77, 90
Vatunjae Discontinuity, 86
Vatupochau Mountain, 11, 17, 88
Vaumandali River, 26–27
Vegetation, 9
Veins, mineralised, 71
Vella Lavella, 92
VERHOOGEN, J., 24–25
Veeru Moli, 46
   Fault, 49, 60, 63, 90
Vindobonian, 39
Viso River, 16, 77
Vitiaz Trench, 1
Volcanic Province, 2, 87, 100
Volcanic wackes, 33
VON NORBEECK, F., 3
Von Wolff diagram, 97
Voza Lavas, 91
Vuraka landslide, 17

Vura syncline (Florida), 92

WAGER, L. R. and DEER, W. A., 97
Wahere Island, 19
Wainimala Group, 91
WALSHAW, R. D., 5, 76
Wanderer Bay, 73, 78
Warahito Lavas, 91
WARDEN, A. J., 92, 97
WARIN, O. N., 74
'Weather Coast', 11, 12
Weather Coast Block, 18, 49
Websterite, 28
WEILAND, E., 74
WEISS, L. E., 48, 52, 59
WELLS, T., 3, 77
WENTWORTH, C. K., 35
WENTWORTH, C. K. and WILLIAMS, H.,
   30, 40
West Melanesian Trench, 1
WESTWOOD, J. V. B., 85, 101
WHEELER, H. E. and MALLORY, S. V.,
   36
WHITMORE, T. C., 9
WHITTEN, E. H. T., 52
WILKINS, R. W. T., 74
WILLIAMS, A., 60
WILLIAMS, H., 29, 33, 40
WILSON, G., 56
WILSON, J. T., 93–94
WINKLER, H. A., 4, 74, 77, 80, 85
WISE, D. U., 93
Woodlark Rise, 1, 101
WOOLLARD, G. P., 4
Wounpouko Calcarenites (New
   Hebrides), 92
WRIGHT, J. B., 93, 101
WRIGHT, P. C., 3, 7–8, 32–33, 42, 58–60

YODER, H. S., 23, 30
YOULES, I., 74

Zinc, 74–75
Zircon, 78
ZOHAR, E., 74–75

Her Majesty's Stationery Office

Government Bookshops

49 High Holborn, London WC1V 6HB
13a Castle Street, Edinburgh EH2 3AR
41 The Hayes, Cardiff CF1 1JW
Brazennose Street, Manchester M60 8AS
Southey House, Wine Street, Bristol BS1 2BQ
258 Broad Street, Birmingham B1 2HE
80 Chichester Street, Belfast BT1 4JY

Government publications
are also available through booksellers

Institute of Geological Sciences

Exhibition Road, London SW7 2DE

Murchison House, West Mains Road,
Edinburgh EH9 3LA

The full range of Institute publications is
displayed and sold at the Institute's Bookshop at
the Geological Museum, Exhibition Road,
London SW7 2DE

The Institute was formed by the incorporation of
the Geological Survey of Great Britain and the
Geological Museum with Overseas Geological
Surveys and is a constituent body of the Natural
Environment Research Council

Printed in England for Her Majesty's Stationery Office
by The Scolar Press, Ilkley

Dd 595767 K10

ISBN 0 11 884080 0*